COMPETE
& WIN in
TELECOM
SALES

By Philip Max Kay

A STEP-BY-STEP GUIDE FOR
SUCCESSFUL SELLING

CRC Press
Taylor & Francis Group
Boca Raton London New York

CRC Press is an imprint of the
Taylor & Francis Group, an **informa** business

CRC Press
Taylor & Francis Group
6000 Broken Sound Parkway NW, Suite 300
Boca Raton, FL 33487-2742

First issued in hardback 2017

ISBN 13: 978-1-138-41229-3 (hbk)
ISBN 13: 978-1-57820-062-7 (pbk)

This book contains information obtained from authentic and highly regarded sources. Reason-able efforts have been made to publish reliable data and information, but the author and publisher cannot assume responsibility for the validity of all materials or the consequences of their use. The authors and publishers have attempted to trace the copyright holders of all material reproduced in this publication and apologize to copyright holders if permission to publish in this form has not been obtained. If any copyright material has not been acknowledged please write and let us know so we may rectify in any future reprint.

Visit the Taylor & Francis Web site at
http://www.taylorandfrancis.com

and the CRC Press Web site at
http://www.crcpress.com

To my wonderful Mom; Anna L. Kay
Three caring sisters; Martha, Renie, and Elizabeth
My best friend and daughter; Debbie
And especially in loving memory of Dad; Milton C. Kay

AUTHOR'S NOTE

Go ahead, try to write a book. Unless you're retired or independently wealthy, there is no good time to sit down and write a book. Certainly not nights, when you're tired after a full day of speaking or training and certainly not weekends. Weekends are for skiing, running, sailing, relaxing or watching the Red Sox.

So it is with great appreciation that I see this project come to an end. For those of you that have completed a project of this magnitude, you know it is only the beginning, because a book does that for you, it gets you thinking about the next, and the next. I would like to thank all the salespeople I've met over the years that encouraged me to put my thoughts together, and especially all the salespeople that have taught me how different selling in telecom was from any other sales job.

I want to thank Christine Kern at CMP for her support, belief in my abilities, and relentless pursuit of excellence, and Janice Reynolds, the editor, for doing very well what she does best. Also many thanks to Claire Belanger at Snow Harbor Graphics for tirelessly making changes to the sale illustrations until we got them right. Lastly, thanks to everyone else that tried at one time or another to vie for my attention when I was writing; from the Yellow Lab "Taffy" to the Pekingese "Patty Mayonnaise," cats/kittens "Poshy" "Skitter" and "Weenie" and especially many thanks to Sally and Ben for the deepest of love and understanding.

Preface

The Telecommunications Industry is the second largest industry in the United States, ranked just after healthcare.

I wrote this book to share with you the techniques that I have developed during my 30 years experience in sales and marketing for the Telecommunications Industry. The secret to my selling success in the Telecommunications Industry is laid out in eight simple, but extremely important steps.

THE EIGHT-STEP SALES PROCESS

STEP 1: PROSPECTING — THE FOUNDATION OF SUCCESS

STEP 2: GREAT FIRST APPOINTMENTS

STEP 3: QUALIFYING

STEP 4: FACT-FINDING

STEP 5: EFFECTIVE PRESENTATIONS

STEP 6: OVERCOMING OBJECTIONS

STEP 7: CLOSING

STEP 8: FOLLOW-UP AND SUPPORT

This book will aid both the veteran and rookie in their goal of achieving a lucrative career in sales for telecom equipment, services and technologies.

If you currently have experience within the telecom industry, the selling process that is detailed in the book will help make your process more effective.

If you don't have experience in the industry, the steps outlined in this book will allow you to bring the best of what you know from having sold in other industries, and customize it to sell telecommunication equipment and services.

My goal is to share with veterans and rookies all the information necessary for obtaining a competitive edge. Each chapter is full of ideas, insights, and information about selling, and specifically about selling in the Telecommunications Industry.

One last note: Selling is a perfect stepping stone for anyone that might want to start his or her own business someday.

One more little item: Please call (978) 526-4200, fax (978)-526-7080, e-mail (Phil@philipmaxkay.com), or leave a message on my homepage (www.philip-maxkay.com) with your questions, your stories, your successes and your failures. If you have any question that you think I can help you with and anything that you feel might be helpful to other salespeople in the Telecommunications Industry, please share it.

I speak frequently to all kinds of groups all over the world, and your comments and concerns will get the appropriate level of exposure. Please use the contact information set out in the previous paragraph.

I assume that if you're taking the time to read this book, then I'm not the first (and probably won't be the last) sales trainer that you will learn from. In fact, I have repeatedly told salespeople to try to take a little from each.

If you learn one thing that is helpful from every sales trainer, think how much better you'll be. So, put this author in the same category as all of the sales trainers that you are familiar with then ask yourself, "What makes this book's advice different from all the rest? What differentiating value does this book offer you?"

First of all, my expertise is strictly in the Telecommunications Industry, and I have a wealth of experience in that business. Here's something even more important, right now there are salespeople all over the country that call me on a regular basis for assistance. They have read this book, or listened to one or more of my audio tape products. They are aware that I know the major manufacturers and am familiar with many of their products and services, and that I conduct competitive sales training seminars, and they are looking for any advantage over their competitor. They know that if I'm doing a seminar or traveling, they can simply leave their name and number and I'll call them back that evening or the next day. Try that with Brian Tracy, Dennis Whateley, Zig Ziglar, Jim Rohn or whoever — that's a value added for you — try it (978)-526-4200.

Contents

Chapter 6 - Step 4: Fact-Finding

Chapter 7 - Step 5: Effective Presentations

Chapter 8 - Step 6: Overcoming Objections

Chapter 9 - Step 7: Closing

Chapter 10 - Step 8: Follow-Up and Support

Chapter 11 - Success and Failure

The Wonderful World of Selling in the Telecommunications Industry

HOW IT ALL STARTED

"So, you want to sell telephones?" Bob Fabbricatore, the President of International Business Telephones (IBT) asked, while looking me right in the eyes.

"Yes, I do" I replied trying to sound as confident as possible.

I was sitting on a soft couch in a magnificent ocean front apartment overlooking all of Boston harbor on a fabulous New England spring day. Bob Fabbricatore and Parker Ladd, IBT's Vice President of Sales and Marketing, waited for my reply.

"Well, I've been doing pretty well selling computers, the market is wide open for telephones, and Norm Peteriet (a fellow salesperson at Burroughs) has convinced me that the future is in phones"

"Good, you're hired," Bob and Parker said simultaneously.

"Great, when do I start?"

"As soon as you can," Bob said, in a matter of fact manner.

"Ok" I responded (not quite sure if I should be excited or depressed), so I continued . . .

"What's the salary?"

"We don't pay salaries."

"What about bonuses?"

"We don't pay bonuses."

"What about expenses?"

"We don't pay expenses."

"What kind of training do I get?"

"We're working on that."

"What kind of benefits do I get?"

"We're working on that too."

"Well, what do I get?" I'm now feeling a little less than excited.

"You get the highest commissions in the industry. a straight 10% on everything you sell. And, of course, a ground floor opportunity to sell phones in a brand new industry."

I kind of felt that if they had asked me to breathe on a mirror and if I had fogged it up ... I would have been hired. The fact that I had a BA in economics and an MBA in marketing, two successful years selling computers for Burroughs (now Unisys), and two years as an officer in the Navy, was quite irrelevant.

I was indeed wearing my favorite, if only, suit (dark blue), and power tie (maroon). Do you think that was influencing them?

Could I sell? That's all that mattered. I could, or at least they thought I could, and the rest is history.

The year was 1973, the telephone interconnect industry was still in its infancy, I was giving up a job selling computers for a Fortune 500 company and accepting a job for a little known interconnect company with a big sounding name: International Business Telephones.

I began my almost 30 year career selling and marketing telecommunications stuff because, after Norm Peteriet joined the revolution happening in the telephone industry, he called me one day at Burroughs with a simple presentation.

"Hey, Phil, I've got some guys you should talk to, Bob Fabbricatore and Parker Ladd, from IBT. You should do it, the future is in telephones not computers"

He was probably half right. It was the beginning of my love affair with the Telecommunications Industry.

SOMEONE HAS TO SELL ALL THIS STUFF

I had so much fun selling in this industry that I've spent the last 15 years travelling all over the world training sales and sales management professionals. I share my sales ideas with brand new rookies, and crusty old salts, alike.

It has been a blast — enthusiastically sharing the excitement of selling — and the Telecommunications Industry has not let me down.

Of course, back in 1973, we were focused on upstaging the incumbent monopoly: AT&T.

The focus was originally on key telephone systems (KTSs), which slowly evolved to private branch exchanges (PBXs). Then call detail recording systems became the rage, and just when the industry needed a quick shot in the arm, the engineers found an application for digitized voice. This, in turn, allowed voice-mail and automated attendant systems (ADSs) to proliferate. And, for awhile, it was a case of the tale wagging the dog, with customers upgrading their phone systems just to get voicemail and ADSs — improving productivity ruled. There was a need for sales people with specialized knowledge, so they could demonstrate the benefits and implementation of products that would allow companies to improve productivity.

At the same time, IBT was not only competing against the incumbents for equipment, but also for long distance services, which they had included in their product line. Now our customers were faced with a myriad of telecommunications usage choices and it was up to someone, who knew their stuff, to help them fight their way through this jungle of new technology.

Computers and communications merged, computer telephony (CT) became the rage, and someone had to identify the applications and show businesses how CT could make their companies more competitive.

Then, just when the customer began to find a trail through this technology maze, the United States Congress passed the Telecom Act of 1996. And hundreds of new, mostly well-funded companies sprang up — going head-to-head with the Regional Bell Operating Companies (RBOC) and Incumbent Local Exchange Carriers (ILEC) to sell local voice and data services. Again, there was a need for someone familiar with local calling services and voice and data bandwidth economies to explain why business should look at these alternatives.

At the same time that Competitive Local Exchange Carriers (CLECs) were installing central office switching and taking on the incumbents in the local calling arena, technology was becoming more and more affordable, PC's were in more than 60% of the homes, and the traffic on the Internet was booming. Again, there was the need for someone who understood telecommunications to explain to all businesses — SOHOs (Small Office/Home Office), SMEs (Small/Medium Enterprises) and large enterprises — how to take advantage of this rapidly changing technology.

It just goes on and on — rapidly changing technologies, a favorable regulatory, judicial and legislative environment, the convergence of computers and communications — and someone has to sell all this stuff.

MY SELLING PHILOSOPHY

Many of the readers are probably familiar with what I do, especially if you sell or have sold in the telecom industry. It is quite possible that the reader has listened to one of my tape sets on selling, or watched one of my sales training videos, or possibly attended one of my live sales training seminars featuring either competitive sales skills or presentation skills, or perhaps my problem solving workshop.

For those readers familiar with my programs, I would like to begin with a short review of my selling philosophies. And for those of you not familiar with my sales philosophies, it is critical that you understand these themes and buy into this premise.

So here goes...the overriding theme is: Prospects in Telecommunications Industry only buy from salespeople they have trust and confidence in… and it helpsif they like you.

"A prospects trust and confidence is essential."

Now certainly that is not such an earth shattering comment. But when you realize that almost everything we do, with regard to interaction with our prospect, either improves their confidence level in us or erodes their confidence in us...then you begin to realize the importance of confidence.

In every chapter of this book, I share with you ideas, selling philosophies, methodologies, and principles that help you endear yourself to your prospects.

THE SALESPERSON MAKES THE DIFFERENCE

Let me begin by sharing with you a short story that dramatically illustrates the importance of the salesperson in the selling cycle.

A friend of mine owned the local pharmacy in the center of the town where I live; the beautiful, picturesque town of Manchester-by-the-Sea, located approximately 20 miles north of Boston, Massachusetts...and....yes...on the sea.

The Allen's Pharmacy Story

Jay was constructing a new building for his business, Allen's Pharmacy, which would be just about one quarter of a mile from where I live. He was doing this mainly to modernize the store, but also he needed more space and convenient parking for his customers.

He knew that I have held a number of sales and marketing positions in the Telecommunications Industry so he asked me about phones. I promptly gave him the names of two of the best salespeople I knew in the Boston area.

Jay contacted both salespeople — each represented a different manufacturer. But (as is typical of most small businessmen), Jay also solicited a quote from the market share leader — Lucent Technologies

I helped Jay with his telecom requirements to ensure that he would provide all three companies with the same set of specifications.

After hearing the sales pitch for each of the three sales representatives, Jay called me at my office. The market share leader — Lucent — just faxed him a proposal; they didn't bother to provide much detail and were eliminated early from serious consideration.

Who is going to sell them their new phone system?

The first salesperson really did a great job of convincing him that his system was the best. Jay told me that when the salesperson left, he was pretty sure that he'd go with that system. However, as it turned out, the second salesperson, to both of our surprise, proposed the same manufacturer's system. This just re-enforced Jay's desire to purchase that particular system. However, the second salesperson took a different tact, according to Jay, and I quote:

"He opened my eyes to a number of other considerations with regard to additional telecommunication requirements and different financing options. He really knew what he was talking about. So I'm going ahead with the second salesperson."

What's the bottom line here? What does this story really illustrate? Two things:

First, how did the second salesperson know there might be additional telecommunication requirements? He asked! What a revelation — *asking, questioning, and probing,* not telling, relating, or talking — is a valuable sales skill that we will explore throughout this book.

SALES TIP:
Focus on asking, questioning, and probing and stop telling, relating, and talking.

Second, Jay's entire conversation with me regarding his decision on how to spend thousands of his hard-earned dollars did not revolve around money or features or service, although I'm sure all were critical factors in the sale. He was talking about the salespeople...yes...the salesperson made the difference...WOW!

> **SALES TIP:**
> In the telecommunications industry, you,
> the salesperson, make the difference.

Throughout the Telecommunications Industry the salesperson makes the difference. It doesn't matter if it's:

- a $50 dollar cellular phone

- a $500 piece of hardware

- a $5,000 key system

- a $500,0000 PBX

- a $5,000,000 central office

- a $5,000 per month long distance contract

- software, services, support

- voice, data, fax, video, or CT applications

Are there other factors? Absolutely! Your product and its ability to fully satisfy the prospects needs are a must.

Your company and their ability to make the prospect feel totally comfortable with the level of service and support is always a crucial factor in the sales decision. In the Telecommunications Industry, however, major decisions are made each day because of you, the salesperson.

Think about it for a minute. This is not necessarily true of other industries.

The Honda Prelude Story

I would like to share with you the story of one of the last automobiles I purchased: a Honda Prelude.

I was looking for a black one so I could put a white racing stripe down the side. Unfortunately, none were in stock, so I told the salesperson I'd come back in a couple of weeks, with the hope that one would come in. The salesperson promptly proceeded to try to sell me a blue one.

Finally, I was able to convince him that I only wanted a black Honda, at which time he stated that he would call if one came in. I informed him that I had a very busy schedule for the next few days, but to please call.

As luck would have it, the very next morning the call comes in at my office. "Mr. Kay, good news, we just got a shipment of Honda Preludes and there is one black one."

"Good," I replied, "I be there on Friday" (two days later).

The response was: "Oh, I'm sorry Mr. Kay, but I don't think it'll still be here then." This is when I started to really dislike that salesperson.

Now I loved that car...I wanted that car...I had to have that car...I liked the smell of that car...I liked the feel of that car...but I did not like the salesperson...I bought that car.

I liked the car, I didn't like the sales person... I still bought the car.

I knew the manufacturer, I knew I could get the car serviced almost anywhere, the salesperson wouldn't negotiate. The salesperson would not even hold the car for a couple of days; so, I certainly didn't have any trust and confidence that he was doing the right thing for me. I thought he was a jerk...he was a non-entity in my purchasing decision, although I did buy the car.

However, in the Telecommunications Industry, you may well be selling a product from a manufacturer that is not a household name. Even if your manufacturer is a household name, you better be on your toes because you'll lose to the competition if you don't sell yourself along with the product.

In this industry, the products are so technically complex that they need to be fully explained in benefit-oriented terms to the prospects. Whether it's a small businessperson who makes this type of decision once every 3-5 years or a telecommunications manager of a huge corporation who makes purchasing decisions involving multiple sites or large sites, all must rely heavily on the vendor salespeople to educate them.

Now, if you're a veteran salesperson, you know this. So it's easy for you to stop reading for a minute and reflect on some of your more recent sales. What was the difference? What really turned the sale toward you, your company, and your product? What did you do or say to influence the buying decision? Stop and reflect for a moment, learn from the self-analysis, and if appropriate, share the revelation with your sales associates.

As a veteran salesperson, hopefully you are reading this book to help you rediscover your excitement in who you are and what you do. This book will help any dedicated salesperson to love selling — and especially selling in the Telecommunications Industry.

I haven't forgotten you rookies — new at selling anything — but particularly new at selling in the Telecommunications Industry...I want to say welcome aboard...welcome to the wonderful world of telecommunications sales.

A CAREER IN SALES

This book's first mission is to remind you why you are or want to be in this business. Have you ever heard the expression: "Love only what you do and do only what you love?" Another way of saying it is: "If you find a job you like you never have to go to work any more."

Due to corporate downsizing or right-sizing — whatever you want to call it — the way we view working for a living in the United States has changed. Everyone reading this book has experienced the effect of this phenomenon either through personal experience, a family member, neighbors, or co-workers.

Today there is no such thing as job-security, at least the job security that our mothers or fathers knew. That is a thing of the past — job security is now an oxymoron...right up there with jumbo shrimp, military intelligence, a legal brief, honest crook, slumber parties, and rap music. Our economy is no longer built upon the success or failure of the Fortune 500 companies; but upon the creation, nurturing, and growth of the thousands of small businesses all over the country.

There is a lot of talk about entrepreneurialism — starting companies, controlling your own future, achieving financial independence, not ever again working for someone else — someone else that can control your life. In fact the word "control" is significant in that when surveyed, a huge percentage of working individuals equated "degree of control over their working life" to be directly proportional to the "degree of happiness in their life." The greater the control the greater the happiness.

Your happiness in life is directly proportional to the control you have over your work.

Do you want control in your life? Do you want financial freedom? Do you want job-security? Some might suggest, especially if you listen to any of the late-night infomercials, that it's time to buy a franchise, and/or start your own business, and join the wonderful world of entrepreneurialism.

I've got a better idea. And you probably know what I'm leading up to, you're right, a career in sales, and, specifically, a career in telecommunications.

Let's put this in perspective. Say you want to someday own your own company. Unfortunately, you don't have the savings, or maybe the know-how, or perhaps you don't have the right idea, product to sell, or you've never borrowed from the bank, or written a business plan or any of the other many subtle barriers to entry as an entrepreneur. It could be that you're just not ready to take the risk — with your future or your family's future. Sales is the perfect stepping stone.

Every small businessperson I know, when asked, can list all the mistakes they made building their business. However, if you're in sales, someone invested in you — they hired you to sell for them. In doing so they took the risk that you could perform for them. Part of their investment in you was the training and education that you needed before you could become an asset to the company.

I want you to stop reading and think about the person that made that first investment, take a few moments to say thank you, [whoever] from the bottom of your heart. Now say it out loud three times...I'm serious — SAY IT.

As you can see I feel very strongly about this. Why? Because a career in selling is all about developing the right attitude...and, if and when, you develop excellent sales skills — the kind of sales skill we'll discuss in this book — you will have all the job security you'll ever need. You will have total control over your life, plus financial security for you and your family. Furthermore, as any entrepreneur can tell you, to succeed in any business requires excellent sales skills. Or to state it another way: The ability to sell is a prerequisite to success as an entrepreneur — although, it might not be in directly selling your company's products and services, it may be in selling your ideas, or raising money.

Why Selling is a Wonderful Profession
I would like for you to put the book down and make a mental note of all the reasons why you're in sales or considering a career in sales — even better — write out a list of all the reasons. I've already discussed a few, list at least five more.

I will now share with you my list of reasons why selling is a wonderful profession.

Job-security: I'm assuming that you've learned the basics of the sales process, then your skills are transferable to other companies. Again, I repeat — selling is a perfect stepping stone for anyone that might want to start their own business someday.

Money: Usually the most popular reason for being in sales is to make money. In fact, if you've read about my background (on the back cover of his book) you'll notice that I've worked for a number of large and small companies in the Telecommunications Industry. In every case the salespeople were the highest paid in the company.

Think about this for a minute: Companies will pay top dollar for people like you to "open doors" because it's the toughest job. But, when learned correctly and properly executed, it can be the most lucrative. It is well-known that sales is the highest paid, hardest-working profession, and the lowest paid non-working profession; i.e., if you're in sales and work hard, you can certainly make a lot of money, especially if you're on straight commission. And if you don't work, guess what, no money. In any event, salespeople usually have direct control over the amount of money they can make.

If you want to be in sales, you need to have an engaging personality, which means:

- You like people.

- People seem to like you.

- You easily can meet and make new friends.

- You are curious about lots of stuff.

- You have a nice way about you.

- People like being around you .

Any of this sound familiar? If so, it's time to realize that companies will pay for you and your talents.

Do you get excited just thinking about how much money you can make? DON'T, I repeat DON'T, until you do the following:

Find the top salesperson in your current company and ask him or her how much they sold last year. That figure should relate directly to how much they earned. Then and only then can you get excited.

Freedom: A career in selling offers the highest degree of independence and freedom. I know, because I have also worked in a number of product development and marketing staff jobs for high tech companies. My co-workers in those positions were extremely jealous of the freedom that the salespeople had — to be at home, to be travelling (on the road), to get their errands done during the day, if need be...then to do paper work or get organized in the evening.

I know many salespeople around the country that are able to balance a successful career in sales and still be home to meet their children when they come home from school. That is not an option for those working 9-5.

Fun: Selling can be a lot of fun but you are the one responsible for making it fun. I've believed all my life that the easiest way to be successful is to work hard and have fun. Most of us lack the self-discipline to continue to work hard year after year at something that is not fun. In fact, if you don't still get a kick out of someone making a check out to you or your company...because of the influence you've directly had on that purchasing decision...then you've probably stopped having fun. This business of selling can be an absolute blast if you develop confidence in yourself, if you really like the company you're working for, and you like your company's products and services.

Minimal politics: I imagine that one of the most depressing situations in life would be to have a job where your direct boss is someone that:

- doesn't like you
- doesn't respect your capabilities
- is jealous of your capabilities
- doesn't appreciate what you do
- never gives you credit for what you do

This can happen, to a degree, in sales, but most often your performance is judged solely on your sales performance. There is nothing subjective in sales, when people buy, they buy. Of course, there are some salespeople who sell on price, cut margins, and their value to the corporation can become questionable, but, on balance, orders are orders...and most corporations measure salespeople on results. I've personally witnessed situations where strong individuals jumped right over a problematic supervisor to establish a lucrative sales career, even though it seemed for a time that they were being stifled by the supervisor or management of the company. In selling "the cream does rise to the surface" and you can't keep a good man or woman down for long. Compared to other professions...this one has minimal career-limiting politics.

Self-satisfying: As you know from my background, I have a BA in economics, and an MBA is marketing; plus almost three years of additional schooling when I went into the navy after college. Although I was almost 26 years old, and had all that schooling, no one had taught me anything about selling. It wasn't until I got my first job selling computers, and attended a two-week selling course, that I discovered what I really wanted to do in life.

There is a giant irony of selling. Ask anyone on the street what words come in to their heads when they are asked to think about a salesperson, and the words are usually not too complimentary.

I will give you an example. I've spent a lot of time training salespeople for major telecommunication corporations, including the entire regional bell operating companies (RBOCs). As their industry becomes more and more competitive, RBOCs are being forced to downsize and cut overhead. In doing so, they have moved many employees from a salaried to a commissioned position. In a number of the RBOCs I visited, the managers wouldn't even say the words "commissions", or "bonuses," instead they referred to them as...listen carefully..."the amount of salary at risk."

For anyone that has been in sales all their life, that is hilarious. A seasoned salesperson regards the commission and/or bonus portion of their compensation plan as their golden opportunity. Since they know full well that if they perform as they are suppose to each week, i.e., prospect and generate activity, find qualified buyers, get appointments, fact-find and qualify, give presentations, do proposals, give demos, overcome objections, ask closing questions, follow-up and get referrals, there is "no salary at risk."

Individuals who have been on salary all their life really fear that they are giving up security, reliability and don't see that if they do their job (assuming the compensation plan is fair), they have the potential to increase their annual income, not lessen it.

Here's the best way I've found to think about selling. If you believe in your ability to figure out what prospects need, and you believe in your company's products and services, then selling is nothing more than "helping people make decisions that are good for them." When you do that, selling is the most self-satisfying job in the world. Remember that every support person in your company, every accounting person, marketing person, manager, operations individual, depends on you and your ability to find new customers. Without you and your customers, they wouldn't have a job.

Career Choices: A recent survey indicated that an unusually high number of top-level business executives came up through the sales and marketing departments of their companies, as opposed to manufacturing, operations, accounting or service. Selling, once mastered, lays the foundation for a myriad of future career choices, not to mention the opportunity to go out on your own.

WHY I LOVE SELLING IN THE TELECOMMUNICATIONS INDUSTRY
Let's look at the many advantages of a career in the wonderful world of telecommunication sales.

What is telecommunications? I understand that the word telecommunications comes from the Latin word, *tele communicatus*...where the *communicatus* means to impart information, i.e., communicate...and the *tele*...means from afar. Telecommunications encompass all the equipment and services — hardware and software —involved in everything from telephones, key systems, PBXs to satellites, microwaves, cable, wiring, and fiber. This includes everything involved in voice, data, image and video communications; local area networks (LANs) and wide area networks (WANs); cellular, wireless and radio; add-ons and value-added products; voicemail and automated attendants; video conferencing; computer telephony products and services; on-line services and databases; direct TV and the Internet.

The industry includes all the manufacturers, distributors and retailers involved in moving those products to market, all the long distance companies and re-sellers of long distance, the local facility-based companies, cable operators, specialized common carriers, and telephone companies...and on and on and on.

Why do I think it's a wonderful world to be selling in?

Once again, put the book down for a minute and make a mental note of why you like this industry or think you might like this industry and why it is such a great industry to be in right now...or if appropriate...write out a list.

THE TELECOMMUNICATIONS INDUSTRY IS THE PLACE TO BE NOW
The favorable regulatory, legislative and judicial environment is one reason the Telecommunications Industry is currently a wonderful world to be involved in.

Without going through a lengthy discussion of all the legislative acts and judicial decisions that have changed our industry, it seems to me that the 1969 Carter Phone Decision was the beginning of real competition in the industry. And about every 3-5 years we have seen major breakthroughs, such as, competition in customer premise equipment and long-distance, the break-up of the Bell system on January 1, 1984, the Cable Act of the early 90's, and the Telecom Act of 1996. All of these events opened up competition in the local calling markets and the government seems to be leaning towards allowing more and more open competition.

Now cable companies are allowed to carry dial tone to homes and businesses, and telcos are allowed to supply video to the home. Therefore, it's inevitable that long distance companies are allowed to sell local lines and the local companies, under increasing pressure to open up their markets, will be selling long distance. With all of this comes many opportunities for salespeople in the Telecommunications Industry.

Rapidly Changing Technologies: You add into this very favorable mix the rapidly changing technologies within the industry and you find multiple windows of opportunity for the salesperson in our business.

First of all, the same changing technologies that are making personal computers more and more affordable are fueling growth in the Telecommunications Industry. We've all watched the phenomenal growth of the Internet, and the growth in the industry due to the convergence of computers and communications. The best place to be in sales is within an industry that's experiencing rapid expansion. Healthy growth is forecast for multiple areas in the Telecommunications Industry.

The changing technologies bring greater productivity tools to users at a lower cost making paybacks and return on investments look even more attractive, which make the telecom products more affordable and easier to sale, since as products become more affordable, greater selling opportunities occur. Anyone that has been in this industry for awhile can think of all kinds of examples, such as, the increasing affordability of automatic call distributors (ACDs), or voice mail systems.

Also the changing technology is forcing the government to re-look at how dial tone, video and interactive programming gets to the house and business. It is the changing technologies that are fueling the advances in personal communication systems, the PC marketplace, and all kinds of companies are positioning themselves to supply the equipment and garner the voice and data revenues in this market alone. Do you think they are going to need people to sell all that stuff?

Confused Customers: Interestingly, it is precisely the rapidly changing technologies, with all the standards combined with deregulation that keeps our potential customers confused enough that they actively seek a strong partnership to guide them through the maze. Therefore, the most successful salespeople in this business in the coming years will be those with the attitude, skills, and knowledge to build strong, lasting partnerships with their customers.

Rapid Growth: The major industry periodicals publish up-to-date statistics on the multiple pockets of growth in the industry. Keep an eye on them — you want to work in an area that is being fueled by growth.

U.S. long distance service revenues are growing at 6% per year and international long distance is growing twice that pace. The competitors to the big three established long distance carriers are experiencing exponential growth as they grab increasing market share. Almost everywhere you look in telecom, you can find pockets of strong growth.

In case I haven't convinced you that high growth rates in the right industry are a big deal, let me relate a short story to help illustrate. My father owned a number of ladies retail shoe stores in the Providence, Rhode Island area. He worked six days a week his entire life. As I was growing up, dad's advice to me was to "get into an industry where you're swimming with the tide, not always trying to swim upstream like us in the shoe business."

My Dad's advice was very sound.

He was in an industry characterized by little growth, fierce foreign competition, volume discounting, seasonal business swings, fickle customers, style changes, thin margins, and worst of all, a weather related environment. All or some of these factors would combine at one time or another to inhibit dad's success.

In telecom, you get a big bang for your buck. Your efforts are rewarded many fold. As companies grow and prosper, individuals that have proven themselves are moved up. Dad was right — it's a lot better to be swimming with the tide.

Telecom products are 100% cost-justified: Telecommunication is recognized as the marketing tool of the future, and as the cost of telecommunications comes down, the cost of telecommunication products continue to be 100% cost-justified through usage savings and/or improved productivity. For example: how would you liked to have been the person selling those automatic scanners to the supermarkets? Do you think that's sped up your friendly market checkout person? How would you like to have been one of the salespeople that sold all those automatic teller machines (ATM's) to the banks? Do you think that has improved the banks bottom lines, improved customer service, sped up cash flow, improved the productivity of tellers and more...of course.

In the coming years there will be all kinds of killer applications in the Telecommunications Industry just screaming for salespeople.

Telecom impacts the bottom line: The products and services appeal to both the top executives of large corporations and owners of small businesses because they have a direct impact on the bottom line. The sales process to the individual employee/owner is a relationship or partnership type of sale and thus we have the opportunity to meet, influence, help and associate with the most influential and exciting men and women in our marketplace.

RE-CAP

So there you have it — why you should have a career in the Telecommunications Industry. Hopefully, having the chance to share why I love selling, and more importantly, why I love selling in the Telecommunications Industry has helped you solidify your rational for persevering in this industry; or helped you decide to enter the wonderful, rewarding world of selling in the telecom industry.

MOTIVATIONAL SECTION

However, remember that nobody ever said it would be easy so I include here my favorite poem entitled "Don't Quit."

Don't Quit

When things go wrong as they sometimes will
when the road your trudging seems mostly uphill
when the funds are low and the debts are high
and you want to smile but you have to sigh;

when care is pressing you down a bit
rest if you must but do not quit;

life is queer with its twists and turns
as everyone of us sometimes learns
and many a failure turns about
when you might have won had you stuck it out;

don't give up though the pace seems slow
you may succeed with another blow;

success is failure turned inside out
the silver tint of the clouds of doubt
and you can never tell how close you are
it may be near when it seems so far;

so stick to the fight when you're hardest hit
it's when things seem the worst that you...
must not quit.

The Eight-Step Sales Process

Before we can discuss the eight step sales process, I would like to share with you a story I heard from Lou Holtz, the well-respected football coach and recognized motivational speaker.

BACK TO BASICS

Almost every time I ask some of the finest sales managers that I've had the pleasure of working with, how they're doing, they say, "ok." When I ask them about the key to their success, they'll emphasize their reliance on doing all the little things correctly, paying attention to details in the sales process, and in particular, not skipping steps in the sales process. Three such sales managers comes to mind immediately:

- Shawn Kane, Claricom (recently acquired by Staples, Inc.), Milford, CT

- Michael Reichert, ESI, Minneapolis, MN

- Ken Holder, Williams Communications, Richmond, VA

All three are tremendously successful in the Telecommunications Industry and they attribute much of their success to hard work and paying attention to the basics.

THE PARROT STORY

Lou relates the following story about a guy that bought a parrot in a pet shop.

The guy goes into this pet shop and sees and hears this incredible talking parrot...and before long with a little bit of persuasion from the salesperson; he parts with $895 dollars and takes the parrot home.

After a few days the guy notices that the parrot isn't talking...so the guy goes back to the salesperson and says...

Hey I bought the parrot a few days ago and he's not talking.

The salesperson says " well... did you buy it a ladder?"

"No I didn't buy it a ladder..."

"You've got to have a ladder in there."

So the guy gives the salesperson another $10 and takes the parrot home.

After a few more days, the guy notices that the parrot still isn't talking so again he takes the parrot back to the pet shop

When he again told the salesperson the problem, the salesperson asked:

"Well did you buy him a swing?"

"No I didn't buy him a swing."

"You need a swing in there"

The guy gave the salesperson another $12, this time for the swing, and took the parrot home.

You guessed it ...after a few days the guy noticed the parrot still wasn't talking...and returned once again to the pet shop...this time when he explained that the parrot still wasn't talking the salesperson said:

"Well, did you buy him a mirror?"

"No, I didn't buy him a mirror." the guy responds now getting a little bit aggravated.

"You need a mirror in there." So the guy gave the salesperson an extra $15 for a mirror and took the parrot home once again with high hopes that the parrot might begin talking...

When he got home the parrot took one look in the mirror, and as the parrot keeled over dead, the parrot could be heard moaning:

"Why didn't you buy me any birdseed?"

What's the moral of the story?

In the end it's the basics that count the most.

In selling, like in caring for pets, you can do a lot of fancy things, but in the end it's the basics that count the most.

Lou Holtz and Vince Lombardi would be the first to tell you that blocking and tackling are the foundation (the basics) of winning football games. How about giving a little food and water (the basics) to the parrot? With regard to selling, how about a little attention to the basics.

THE SALES CYCLE

Let's begin by looking at the sales cycle. I like to think of the sales cycle as the things that must happen between the time you're sitting at your desk contemplating what to do to the time that you receive a check from a satisfied customer.

Just like we did in Chapter 1, I want you to put the book down and verbalize out loud your steps in the sales process.

> **SALES MANAGER'S EXERCISE:**
> Have your salespeople or a prospective salesperson write out the steps in the sales cycle. You'll learn a lot about what they know about selling, especially if they miss one or two critical steps.

Now lets see how your steps match up to mine.

I will set out my steps first and then I'll go back and explain. If you're a salesperson, with a quota, and primarily judged by how close you are to, or how much you are above your quota, then your life should revolve around the following eight steps to success. If not, and you're spending your time doing other assignments, then maybe you'd better question why, or at least start requesting your company to provide you additional payment for the technically "non-sales" services rendered.

This might seem like a strange comment to you. But remember, I had an MBA in marketing before I ever sold my first telephone system. My company knew that, and every once in a while I was asked to complete some marketing oriented assignments. I usually happily accepted these assignments, only to realize much later that my success or failure was judged solely on what I sold, not on my fabulous marketing skills.

THE EIGHT-STEP SALES PROCESS

Step 1 Prospecting — The Foundation of Success

Step 2 Great First Appointments

Step 3 Qualifying

Step 4 Fact-Finding

Step 5 Effective Presentations

Step 6 Overcoming Objections

Step 7 Closing

Step 8 Follow-Up and Support

How closely did your steps match mine? Did you bunch a couple of the steps together like I did, or did you put them in different order?

OTHER SALES PROCESSES

One of the finest sales trainers that I've ever been exposed to, William T. "Bill" Brooks, uses a clever acronym to get salespeople to remember his six steps to success. Bill says that with his process salespeople are reinforced to not skip any steps and to stay on track.

His six steps are built around the acronym, "IMPACT", which stands for Investigate, Meet, Probe, Apply, Convince, and Tie it down.

- Rather than calling it prospecting, Bill calls it "investigate"...since you need to develop a detective-type of attitude to find qualified buyers for your product and service.

- Rather than talking about what to do in the first appointment, Bill calls it "Meet."

- Rather than call it qualifying, fact-finding or a needs analysis, Bill calls the next step simply "Probe."

- Rather than calling it demos, proposals, presentations, Bill calls it "Apply."

- Rather than call it overcoming objections and closing, Bill calls it "Convince."

- Rather than follow-up and support, Bill refers to it as "Tie it down."

It a good way to remember the steps and I'll share with you a few more of Bill's secrets as we go along.

Since I have worked with most of the major telecommunications manufacturers and distributors over the years, I can tell you that nearly all of the companies follow some sort of standard sales process. How they refer to the process, the words they use, usually depend on the emphasis on that particular segment and the way the process is taught.

One major company I worked with used this:

Qualify

Earn the right

Conduct the needs analysis

Determine the buying process

Propose solution

Handle objections and close

Implement solution

Some of you might be familiar with another sales process that uses the simple acronym of SAVE:

Survey

Analyze

Validate

Execute

Plus I have seen other simple sales processes set forth, such as:

Find the problem

Find the cause

Show the solution

Explain the benefit and close

Maybe you have your own favorite way of remembering the sales process. It is almost guaranteed that you will find whatever current sales process you are using buried within the sales process I outline in this book; but mine is **<u>specific</u>** to the Telecommunications Industry.

All of the various sales processes I have mentioned are just another way of looking at the same process. Although couched in other "words" the concepts are the same as the eight-step process described in this book.

CONCLUSION

If you have experience within the telecom industry already, the process that follows should help make you more effective.

If you don't have experience in the industry, what follows will allow you to bring the best of what you know from having sold in other industries, and customize it to sell telecommunication equipment and services.

If you've never sold, what follows are the eight steps to selling success in the Telecommunications Industry.

Step 1: Prospecting — The Foundation of Success

This chapter is all about prospecting — specifically all about prospecting for success in the Telecommunications Industry. It is the first step in the eight-step selling process.

DEFINITION OF PROSPECTING

Prospecting is simply the process of identifying those businesses that might benefit from using your product or service. Rest assured there is no question about this — it is the most important ingredient in reaching and maintaining a high level of success.

THE EDDIE SAMP STORY

Why does a president of a successful company still have to prospect? In answer to this question, I'd like to begin with short story about prospecting.

A longtime friend of mine, Eddie Samp runs a highly successful mail order computer hardware and software company. I first met Eddie when the two of us were the top two salespeople for Burroughs Corporation, before Burroughs merged with Sperry-Univac to become Unisys.

Later, Eddie sold mainframe computers for Honeywell, then went to Harvard Business School, before starting his own company.

I went to visit Eddie, sat in his beautiful office of his $10 million dollar company and while there noticed on the credenza behind him a computer print out and asked him about it. I was curious because the rest of his office was neat...you know the type of executive's office where everything is in order and the only file out is the one they're working on. Eddie told me it was his prospect list. I was quite surprised — why did the president of the company have to prospect?

Eddie started talking about the financial state of his company with regard to future equity investments. He pointed out that even though he did not need more capital at the moment, he knew that sooner or later he would, particularly if he wanted to continue to grow the company. He had a prospect list of potential venture capital people, bankers, investors and such and was calling now to introduce himself...yup...cold calls.

I wanted to share that story, because this chapter addresses the hardest part of your job — prospecting — and, if the truth be known, the part most of you hate the most. But if prospecting is mastered, it is a skill that (no matter what level you attain in the business world) — for example, Eddie Samp — that will continue to play an important role in your chosen career.

PROSPECTING: A NUMBERS GAME

Many of you are working for a sales manager that manages "by the book" so to say. These sales managers know that a big part of selling is a numbers game. They've been around the industry, and possibly around your company long enough to know that many of you are stuck at a 40% closing ratio. They know the formula for success is simply the number of qualified prospects, times the closing ratio of 40% equals sales. There is a lot of merit in this. If you want to close 4 sales, for example, you need to be actively pursuing sales with 10 prospects. That's fine, but what happens when you make your quota of 4? Do you think the company is going to give you a quota of 4 the next year? Of course not — the next year your quota is going to be 6, so if you assume the same closing ratio of 40%, then you're going to need to be actively selling to 15 prospects next year. Guess what — if you make your quota of 6, your quota the next year will be 8, which means that at a closing ratio of 40%, you're now going to have to have 20 prospects, and on and on.

SALES TIP:
Number of opportunities times your closing ratio equals your total sales.

What all this means is that if you have a consistent closing ratio, then to be successful year in and year out and to meet the rising sale quotas, you must be a diligent prospector.

Later I will share with you the selling skills to improve your closing ratio, but at the moment let's look at what you can do to improve the other part of the equation — improving your ability to find more qualified prospects.

Let's look at prospecting as a numbers game. The more calls you make the better your chances of finding a potential customer. Beginning salespeople will have to start out with high numbers, but as you work on your sales script, presentation, your lead/networking capability and ability to qualify, your numbers should come down

To be successful at prospecting you have to set some goals. The following numbers are just guidelines for the telecom industry. If you're in the equipment end of the business, I suggest a minimum of 25-40 new contacts per day. Set a goal and track your progress.

Here is a guideline for prospecting:

- 25-40 contacts per day (telephone calls, cold calls, call-ins, and leads)
- 10 new qualified prospects per week
- 3-5 presentations/demonstrations per week
- 1-2 closes per week

Getting Computerized

Let's get specific here — I know that many of you are highly computerized. It definitely was easy for me to buy and install the Act contact management software on my own computer, and I have entered over 3000 contacts into the database.

If you're not computerized, then it is imperative that your sales manager share with you the system used by the sales staff, since you must keep track of your calls and callbacks, along with the information you gather and track on your prospects.

Act is just one of many contact management software programs available, just go down and check out your local computer store. However, Act does work great for me, and I love it.

By the way, the telephone number for customer service for my Act contact management software is 1-800-441-7234 and you can obtain ordering information from 1-800-568-9501.

Develop an Attitude

So now that you are organized, let's work on your attitude. I want you to develop an attitude, so before I go any further, I'd like you to put the book down for a minute and think for a minute what I mean by "an attitude." Okay, now let's move on — here's what I mean:

First of all there is nothing you'll do in sales where the potential for rejection is higher than when you are making cold calls.

Rejection...Rejection...Rejection

There is a formula for handling daily rejection. Assume that each sale is worth 100 points. Then assume that you need 5 qualified prospects for each closed sale. This means that day in and day out you're going to get 4 *Nos* for every 1 Yes — *no no no no* yes *no no no no* yes *no no no no* yes — that is the facts of life for a salesperson, so get use to it. And DON'T think of a yes as 100 points, BUT think of each *no* as a step toward the yes, therefore each *no* is worth 25 points.

Next you must learn to change your attitude toward failure and rejection.

Think of failure as a learning experience, i.e., what can you take away from the negative experience and how can you benefit from it.

Think of failure as the negative feedback that might be needed for you to change direction.

The Homing Torpedoes Story: I was an officer on a destroyer in Vietnam. One of our duties was to find, track, and, if needed, attack enemy submarines. Obviously the subs weren't Vietcong subs, but Russian that were in the Gulf of Tonkin watching what our carriers were doing. Our job was to protect the carriers from the subs. To aid us, we had on board a number of active homing torpedoes. Once launched these torpedoes didn't just go straight like the ones you see in the old WWII movies; they sent out a sonar signal, and received a return signal, if the contact moved the torpedo would change direction so as to continue on a straight path to enemy contact.

See the analogy? What can you do differently? Do you change course if the target requires you to? Do you know how to change to be more effective?

Think of yourself as an active homing torpedo.

Think of failure or rejection as an opportunity to develop your sense of humor. If you have been rejected recently, put the book down and think of something funny about that experience. Don't underestimate a good sense of humor in this business.

> **SALES MANAGER'S EXERCISE:**
> At your next sales meeting find the salesperson that has most recently lost an order to the competition....give him or her a good razing...and don't stop until that salesperson comes up with something funny. Demonstrate the importance of a sense of humor in the sales process.

Think of failure as the opportunity to practice your sales presentation and techniques. Every successful person I know has experienced failure in his or her life. One of the best quotes I know is about persistency:

Persistence

Nothing in the world can take the place of persistency
    Talent will not;
Nothing is more common than unsuccessful men with talent

    Genius will not;
Unrewarded genius is almost a proverb
    Education alone will not: the world is full of educated derelicts
Persistence and determination alone are omnipotent

Persistence and **Determination** are fighting words and quite contagious words.

Finally, think of failure as the game that you must play to win. Here's an example:

Baseball fans all know what the Cy Young award is — it's given in each league to the most outstanding pitcher in baseball and named after the great Cy Young who holds the record for the most victories by a pitcher. What most people don't know is that Cy Young also holds the record for the most losses.

TYPES OF PROSPECTING

It's time to look specifically at the various types of prospecting.

Warm Calling

Some call it cold calling or hot knocks...or a bunch of other things...I prefer to call it warm calling. Warm calling is cold calling armed with information and enthusiasm.

You are way ahead of the game if, before you make the call, you have researched and obtained at least some kind of information on the account:

- The type of company.

- The name of the decision maker.

- The prospect's industry.

- The prospect's major competitor.

- Something interesting or newsworthy about that prospect's industry.

Salespeople have asked me why information is so important. I will give you an example, from my own personal experience.

I work for myself, and prefer to answer the phone myself when I'm in the office. Therefore, I have the opportunity to talk to a lot of salespeople. If that salesperson knows a little about me, and shares that with me when they introduce themselves to me, it dramatically increases their chance of getting an audience with me.

Many salespeople have asked me what I think of the different types of prospecting, primarily the advantages and disadvantages of physically driving around and calling on companies verses making telephone calls. It should be a combination. There are many benefits, if you're selling equipment, to visiting an account — you can get a feel for the quality of the building, neighborhood, type of employees. Also, many times you can actually see the most important telephone in the account — the operator's console.

On the other hand, by making telephone calls, you can cover a lot of ground, so to speak, in a relatively short period of time.

SALES TIP:
For those of you that are already selling telecommunications equipment, primarily telephone systems, there is a sure way to determine if you are fully committed to the industry. When driving around, salespeople in this industry tend to develop a weird sense of the world, and specifically their sales territory. For example, when I was still a novice in the industry, I drove around with a veteran salesperson that looked at a building, and said, "Wow, there must be about 50 of 'em in there." The first time I heard it, I wasn't sure what he meant — "50 phones of course."

What I like to do is combine an actual visit with a follow-up phone call. You can visit the account and get the name of the decision-maker, and find out a few things about the existing phone system, telecommunications usage, phone service or possibly any plan for relocation, leave some information and then follow-up with a telephone call to make an appointment.

Or, if you make a telephone call to a prospective account you might be able to get enough information during the call to then follow-up with an actual visit. Armed with the information obtained over the telephone, you then have a better chance of getting in to see the decision-maker.

When visiting an account for the first time you must have a game plan, you must know exactly what you're trying to accomplish. Put this book down for a minute and make a list of appropriate questions you should be asking yourself before making contact.

SALES MANAGER'S EXERCISE:

Ask your salespeople, or a prospective salesperson, to make a list of questions they should ask themselves before making a cold call on an account in their territory. You'll learn a lot about how well prepared your salespeople are, or a lot about the professionalism of a new hire.

I suggest the following questions should be asked before contact with a prospect:

1. What kind of business is this?

2. What do you know about this business from the exterior of the office?

3. What types of communication applications are they likely to have in this office?

4. What type of reception area can you expect on the other side of the door and what type of greeting should you expect when you open the door?

5. What is your opening statement going to be?

6. What is the goal of this particular contact?

7. What questions will you ask?

8. What will you leave (if anything)?

9. What will you do and say if you're invited in to see a decision-maker?

10. What do you anticipate to be your next step at the completion of this initial contact?

After the initial contact, you also should ask yourself some additional questions. So, put down the book and make your list. Now, look at my list of questions you should ask yourself after contact:

1. Did you accomplish your goal?

2. What is your next step with this prospect?

3. Should you follow-up with a thank you note?

4. Can the information you gathered be of use on future contacts?

5. Do you have a product that will meet their needs?

6. What is that specific product and why that particular one?

7. How did that prospect view you as a salesperson?

Can you think of any more?

When calling an account that you believe might need a new telephone system or might be moving...ask for the customer service department. These people are trained to "help." When talking with them, be honest — tell them you need their help — who you are, and ask them who is the decision-maker in that specific situation. Also ask what do they like best about the existing system and what would they change or modify if they were to get a new system. Armed with this information you'll be way ahead and have a much easier time talking to the decision-maker.

For those of you that are selling telecommunications hardware, in many cases you are looking for factors that would influence a telephone system purchase. Think of the factors and make a list. Here's my list of the factors that would prompt a telephone system purchase.

1. Moving

2. Organizational change

3. To save money

4. To make money

5. For new technology

6. To improve worker productivity or efficiency

7. To save energy

8. To keep up with the competition

9. For new features (IVR, ACD)

10. To improve capabilities (voice mail, automated attendant, CTI)

11. For expansion or contraction

12. Out of capacity

13. End of lease agreement

Telephone Prospecting

Face it, if you're going to succeed in sales you've got to become great on the tele-phone...make it your friend...if you spend a lot of time on the phone you might seriously consider a comfortable headset. My contact management system auto-matically dials the number for me...WOW!

Your success will depend equally on the following, and only the following, three variables:

- Your list (accuracy and thoroughness).

- Your script (what you say).

- Your enthusiasm or attitude (how you say it).

Start with the most up-to-date, highly targeted and focused list possible. Then, as mentioned earlier, the more information you have about the account prior to the call, the easier your job will be in terms of establishing a rapport, getting to the decision-maker, and getting an appointment.

Your telephone sales success depends upon your list, your script, and your enthusiasm.

The obvious advantage of telephone prospecting is that it saves time and energy, plus it allows you to make numerous contacts in a short period of time.

SALES TIP:
It is as possible to sense a smile over the phone as it is easy for a prospect to detect boredom.

The First Telephone Call: The subject of what to say and how aggressive to be on the first telephone call has become quite controversial in the Telecommunications Industry. I know that if you are selling hardware, the first call should only be to either sell the appointment, or (if you typically sell equipment over the phone) simply to build rapport, qualify, and gather information.

Many customers I've talked with who know I train salespeople have complained bitterly to me about the aggressiveness of long distance salespeople. We're all familiar with the backlash against "slamming" in the industry. Recently, while working with a number of sales forces selling local services, I noted that these salespeople would do anything to position themselves in a different arena than long distance salespeople. Such as, all stressed the word "local" in their approach.

The most successful long distance and local telecommunications salespeople I know, use a relatively low key, multi-call approach to the customer. The first call is to build rapport, qualify, and gather information. They then send information to the customer, to build credibility in their company, then they ask for copies of bills to analyze. Many companies suggest a survey type call as the first call, then a follow-up call.

Many hardware providers are also selling local and long distance services. Remember that most of the local telephone companies have tremendous credibility with their customers. This is not usually because their service and support are so great (indeed, they're quite vulnerable in these areas), but because of their long-term commitment to the marketplace.

Rules for successful telemarketing:

1. Call your prospect by name (if in doubt always err on the side of adding Mr. Or Ms, to the last name...people love to hear the sound of their own name).

2. Always ask for help (people love to help people).

3. Establish a productive, motivated state of confidence before calling. (Do this by relaxing and thinking of a positive experience related to your product, company, or the sales process.)

4. Do not take rejection personally.

5. Remember again, if you are an outside salesperson, never sell anything over the telephone, you are simply selling the appointment.

6. Lastly, a good telephone solicitation, should be a telephone conversation. (Give your prospect time to talk). After you say, " how are you today," pause — the response will tell you if the person you're talking to is in a good, receptive, or belligerent mood.)

About scripts - If you sell local loops, or local dedicated telecom services, you might want to lead with something like this:

Sample telephone script #1 for getting an appointment: "Hi I'm Phil Kay with XYZ Company, an authorized agent for XYZ RBOC. I would like to send you information on some recently developed telecom options available with your telephone system. So that I may send the proper information, I need to know a little bit about your existing system, like the number of lines and phones you currently have, and who is the appropriate person to send this information to."

Depending upon the friendliness and willingness to help, you can keep probing at this point. If the system is within the competitive window of your company's product offerings, then you can go there for an appointment or maybe hang up and call back later for an appointment.

Here is absolutely the most direct script I've heard for setting an appointment in the telecom industry after you get the decision-maker on the phone.

Sample script #2 for getting an appointment: Hi, I'm Phil Kay from XYZ Company (then add a tag line, such as an authorized agent for the RBOC), or (the largest XYZ type company in Detroit), or (a regional telecommunications company with over 1000 customers here in Austin). The purpose of my call is to set up an appointment to see you briefly. I would like to share with you a few ideas that will enable you to increase revenue or reduce expenses through telecommunications. I promise to be brief...which day is best for you?

You can use these versions as models, or adapt them to fit your specific needs. But whatever you do, develop and write out a script, memorize it, practice it and deliver it — to ensure that it comes out natural, not canned.

Many salespeople hesitate when a sales trainer tells them to memorize something, they think it might come out canned. My answer to this is to think of Dustin Hoffman in the movie Rain Man — if you saw the movie you know that after awhile it didn't even seem like he was acting — although we all knew that he had memorized his lines — he seemed so natural.

Think about Tom Hanks in the movie, Forrest Gump. Tom Hanks seemed so polished, so in character, that he fully took on the character's personality. Tom had memorized his script so well that he could internalize the thoughts and feelings of the character.

For some salespeople selling is indeed a little like acting. This doesn't mean that to be successful you have to be something other than yourself, quite to the contrary. It is critically important to be yourself, to be believable, and most important, to be natural. That is the reason for scripts and preparation. Only if you're prepared and know what you are going to say, will you say it smooth — like an actor.

SALES TIP:

Some hints for successful telephone prospecting:

Be charming without being overbearing

Be serious without being dull

Be forceful without being obnoxious

Be logical without being technical

Be cordial without being cute

Cold Calling or Foot Canvassing

I've shared with you an opening script for a telephone prospecting call; equally important is to develop and practice the opening words of your very first visit with an account. My words can be adapted to your style...but make sure you prepare something...maybe it's trite to say, but; — you have one chance to make a good first impression.

Initial Benefits Statement: Some guidelines for a powerful initial benefits statement (IBS) (some trainers might call it a value statement) might be:

- It must be benefit oriented.

- You should be humble.

- It should be reference oriented.

- Telecommunications should be mentioned.

Initial benefit statement script for first meeting: "Good morning Mr. Jones, thank you for inviting me in. Sometimes as a result of a visit like this we can dramatically increase revenues or decrease expense related to telecommunications. I'm not sure if we can do it for you and XYZ Company but were doing it for (number) customers in the (local area)"

Here's an alternative:

"But were doing it for four other prominent law firms in the downtown Boston area."

You can continue with:

"To determine if we can help, I'd like to ask you a few questions about your company, share with you a little about my company, then we'll decide where we go from there. Does that sound ok?

Set Goals: I suggest a number of goals for foot canvassing, ranging from the absolute best outcome possible, to a fallback position that is a reasonably good use of your time.

1. Close the opportunity.

2. Have an appointment with decision-maker.

3. Set an appointment with decision-maker.

4. Get information on company, decision-maker, telecommunications information, and a return ticket.

There are certain products and services in our industry where there is a reasonably good chance of a close on the first visit. If you fit into this category, then walk in with the expectation of a close. One example is the salespeople that sell yellow page advertising. Many of those salespeople are considered "outside" salespeople. The advantages they have are the longevity of their product and a proven track record of results. They compete, however, against other uses for their customer's advertising dollars.

The next best alternative after a close would be to have an appointment with a decision-maker. If you can't accomplish that on your cold call, then try to at least make an appointment to see the decision-maker at a later date. Lastly, if you've been unsuccessful at any of the above goals for your visit, then at least leave with information on the company, the decision-maker, the number of lines coming into the company (if appropriate), number of phones (again if appropriate), and a return ticket.

Handling the gatekeeper: When you are cold calling in your territory, you are going to walk into buildings and reception areas and meet with what we commonly refer to as the "gatekeeper." In many cases this is the receptionist, sometimes, especially in small to medium size businesses, also the telephone operator. In almost all cases, he or she is also responsible for shielding the owner or decision-maker in the company from unwarranted interruptions from salespeople.

When you walk in the gatekeeper usually knows why you're there. For that reason, what you say, how you say it, how you look, and your demeanor are important. You should know exactly what you are trying to accomplish (your goals). Believe me, if you don't know what that is, it will never happen.

Don't just drift in — but with your goal or goals firmly in mind, walk in like you mean business. Always know exactly what you are going to say, and smile. Here are some suggestions for handling the gatekeeper:

- Do look like you mean business.

- Do look like you're in a hurry.

- Don't just drift in and hang around.

- Don't pretend you're there for something else.

- Do realize that they know why you're there.

- Do know exactly what you are going to say.

- Do treat the gatekeeper with respect.

- Do smile.

Combination: Cold Calling and Telephone
The third type of prospecting is a combination of cold calling and telephone prospecting as described earlier. It can be done in a combination of calling the prospect, then visiting them, or you can first visit the prospect and then follow-up with a call. In any event, warm calling, armed with information about whom you are talking to or meeting with and a show of enthusiasm for how you can help them, is preferable to cold calling.

Referrals
A good referral is better than any other type of prospecting. You've already overcome the credibility issue. Armed with a lead, it is much easier to establish rapport and set an appointment. Just think for a minute how nice this referral script sounds:

"Hello Mr. Jones, and thank you for taking my call. Yesterday, I was talking to your friend, Mr. Stevenson. Recently he signed up for our long distance services. He told me about your company, and encouraged me to give you a call to see if I can help you as well."

When you make a sale or have a conversation with an existing customer or simply talking to friends, always ask if they know someone that would benefit from your product or service. When you call you can use their name. One bold sales trainer I know even suggests that you have your friend or customer, call the contact for you, and introduce you to the new prospect.

Lead Networking

For those of you that sell equipment, the best way for you to think of lead networking is as everything that happens when a company moves. In the equipment end of telecommunications, salespeople should always be searching for companies that have an impending event. An event like a move to a new location is probably the most dramatic example of an impending event. Here is a list of the types of companies that could get involved when a company moves:

Contractors
Real estate agents
Construction companies
Rug companies
Electrical contractors
Plumbers
Window companies
Flower companies
Moving companies
Furniture companies
Interior decorators
Carpenters
Coffee companies
Computer companies
Sign companies
Stationary companies
Printers
Advertising agencies

Many of the most successful salespeople I have worked with in the telecom industry have told me that a great deal of their success has been achieved due to their contacts with individuals working in one or more of the above job categories and or companies.

Lead Clubs

Lead clubs can be a huge source of leads in the telecom business. I would suggest that you put together just a small group of three or four individuals — each a salesperson in a different industry. You can meet every other week for break-fast and share leads. If someone doesn't bring leads, or is not participating, then look for another individual.

Of course, you are going to have to bring leads to the group meetings in order to benefit from the others' leads. The beauty of this is that it will encourage you to prospect. Quite often a company might not be a good prospect for your products or services, but just might be a terrific prospect for one of the members of your group.

RE-CAP

You must take a long-term view of your career. It takes a long time to establish a relationship that has the trust so that you are recommended to that person's best clients. Why? The reasons are obvious, many salespeople have already invested a lot of time building their own relationships and it is necessary that they get to know and trust you before recommending you.

If you start now, however, before long your network will begin working for you. To begin, join your local chamber of commerce or any other local business groups. Become active and volunteer for assignments. Become a tenacious col-lector of business cards, call these contacts often, leave voicemails, socialize with your friends, talk business, make sure they all know who you work for and what you do. Always show your pride in your company, its products and services, plus make it clear you enjoy helping people.

Step 2: Great First Appointments

In this chapter you will focus exclusively on the fine art of the first meeting or appointment with your prospects. As stated previously, you only get one chance to make a good first impression. If you fail, everything else that happens afterwards in the sales cycle will suffer.

REVIEW

First you need to review briefly what you covered in Chapters 2 and 3.

Chapter 3 covered prospecting and getting appointments in the Telecommunications Industry. I want to caution you that it will do you absolutely no good to read this chapter if you're not skilled in finding your prospects.

In Chapter 2 you reviewed the sales cycle, or what I call the eight steps to success, which are:

1. Prospecting - The Foundation for Success

2. Great First Appointments

3. Qualifying

4. Fact-Finding

5. Effective Presentations

6. Overcoming Objections

7. Closing

8. Follow-Up and Support

You can easily see that once you are skilled in finding prospects, you can then use the skills covered in this Chapter to dazzling them in the first meeting.

As I begin to discuss the fine art of meeting face to face with your prospects, please make sure you are comfortable with the concepts we covered in Chapter 3.

SALES MANAGER'S TIP:
Ask your salespeople what they learned or remember from Chapter 3 before going on to Chapter 4.

In Chapter 3, among other things you discovered —

The importance of developing great prospecting skills:

1. Prospecting by the numbers.

2. How to automate the prospecting process.

3. Tools you can use to contact and follow-up effectively.

4. How to develop the right attitude toward prospecting.

5. How to deal with daily rejection.

You read about the different types of prospecting: cold calling and telemarketing and the advantages and disadvantages of each...and my favorite technique — warm calling.

We reviewed the questions to ask yourself before making contact and the question's to ask yourself after making initial contact.

I also covered the factors that would prompt someone to want a new telecommunications system or prompt them to desire new telecommunications services because, only if you know the factors, can you then determine the questions to ask to uncover the need.

You analyzed the three key ingredients to successful telephone prospecting: your list, your script, and your attitude and learned how important each is to your success.

You looked at ways to handle the gatekeeper.

Then you read about the art of networking and getting referrals. Also, once a prospect is found, I demonstrated how you could develop a powerful, compelling, benefits oriented "initial benefit statement" so as to get a prospect's attention.

The Weapons Officer Story

As mentioned earlier in this book, after I graduated Colby College, in Waterville, maine, I joined the Navy as an officer, and soon after boarded my ship, a Destroyer, named the USS PRESTON (DD-795). I'd like to share with you my "weapons officer story," as an analogy to my stepped sales process.

One of my best friends on the ship was the Weapons officer. He would walk around the ship subtly reminding people that his job was the most important job on the ship because without his awesome weapons capabilities we were just a floating platform. As a member of the engineering department, I had to constantly remind him that without the ship's four boilers producing steam for its two main turbine engines, along with the able assistance of the navigators to put his weapons in position to fire, his weapons were useless.

Just like in the Navy where all the different components must work together and build upon each other for a fighting ship to perform effectively, so, too, is it necessary to master all the steps in the sales process to sell effectively.

OBJECTIVES

I should point out here that I've probably used the word "prospecting" a little too loosely; and up until this stage, especially before you've even had a chance to meet with these people, they should only be referred to as "suspects."

Only after the first meeting and putting into practice some of the skills you are about to learn should you begin to refer to them as prospects, or prospective customers.

So now that you've found someone that will talk to you about new telecommunication's services or systems and has agreed to meet with you in person, I want to now work with you on perfecting your first appointment with your "suspect."

Before I share with you my list, I'd like you to put the book down and write out a list of what you think are the objectives you should accomplish during the first appointment with a prospect/suspect.

SALES MANAGER'S EXERCISE:

Have your salespeople, or a prospective salesperson, write down a list of what they want to try to accomplish in the first meeting with a prospect. Compare their list with the following list. You'll learn a lot about your salespeople or a lot about someone's chances of becoming a successful salesperson.

The objectives for a first meeting with a "suspect" should include some, if not all, of the following:

1. To establish a partnership type of rapport with the prospect.

2. To search for commonalties between you and the prospect and between your company and the prospect's company.

3. To set up a timeframe of activities.

4. To qualify in all aspects of the sale, such as for the decision-maker, financial capability and budgeting, impending events, or variables that would influence a decision.

5. To begin to gain the trust and confidence of the prospect.

6. To position yourself for the long-term.

7. To establish your credibility.

8. To establish credibility in your company, products, and services.

9. To be so prepared for the first appointment that you are in control at all times.

10. Lastly, if qualified, get the prospect's commitment to move to the next step.

All of the above should be done so as to establish a helpful, friendly, and especially non-hostile and relaxed environment for the meeting.

There are many different approaches you can take to accomplish the above objectives. Certainly if you and your sales manager agree to additional objectives, include them in your list. It is important (whether you are a veteran salesperson or a rookie) that you develop your own list at the conclusion of this Chapter. That way, no matter whatever technique or style you develop in your first meeting, at least you and your sales manager will know if you've accomplished your objectives.

As someone that has studied sales over the years and has been exposed to the best sales trainers in our industry, I've come to the following three conclusions about the process I am about to present to you.

1. Have a script, memorize it, and it will come out natural.

2. Be wary of applying generic sales training principles to the telecom industry.

3. Decisions in the telecom industry are based solely on product, price, and service.

SCRIPTS

As mentioned in the proceeding Chapter, you should have a script prepared for your first appointment. You should memorize it, then and only then, will it come out naturally. Many of the major telecommunication companies that I have worked with do a pretty good job of presenting their salespeople with powerful and compelling company stories. How you, as the salesperson, internalize the story and use it with your client is what this is about.

As stated earlier, the best way to think of this is to think about Dustin Hoffman in the Movie, "The Rainman" or Tom Hanks in "Forrest Gump," where, after awhile, you don't even realize that they are acting. I'm not inferring that sales-people should act; but if you know where you are going with your discussion, then you can be relaxed, in control and most important, you can be yourself.

I'd like to reinforce a point: I don't want to change anyone's selling "style." Many times it is your individual "style," or your "uniqueness" that is an important part of your success. I only want to give you a framework to work within. Although, I will be sharing with you a proven-effective script for the first meeting, you will want to deliver the script in your own words, and with your own sense of style.

As I travel around, principally in the United States and Canada, it is obvious that different areas of North America have their own unique selling characteris-tics. My heavy Yankee accent is well received here in the Northeast. But, unless you're from Boston, you probably would not be able to replicate my Yankee accent. Likewise, I would feel awkward trying to use the term "you all".

Take the scripts that follow, being careful not to change the meaning of the words, and amend them to fit your unique culture, style, and selling personality.

BE WARY OF GENERIC SALES TRAINING

The following first appointment scenario is specific to the Telecommunications Industry. I personally have sold millions of dollars worth of telecommunications equipment and services over the years and these approaches worked great. You should be very careful taking generic sales training stuff and trying to apply it to the telecom business.

But, in the same vein, I know that a great deal of this would apply to selling other stuff because I'm asked all the time if I train in other industries.

Likewise, I am familiar with the training of the most popular sales trainers, and while many of them will inspire and motivate you, a word of caution should apply. Some of the training, for example, applies specifically to retail sales, and some to short term relationships. Some of the training is specifically for telephone salespeople, and some is industry specific, such as for the real estate industry. I am convinced of the uniqueness of the "sales process in the Telecommunications Industry." If you qualify too heavily early on, you could lose your effectiveness; if you attempt to fact-find before you have established a sound rapport, you will get garbage; and if you try to close before they are ready, you will be alienated.

Many of you are fans of the finest motivational speakers and sales trainers, and

I, too, have my favorites. Attend every session you can, and purchase and read the works of as many as possible. If you pick up a few powerful ideas and tips from each, you'll be far better off. My caveat is simply to caution you from direct application to the Telecommunications Industry.

DECISIONS ARE BASED ON PRODUCT, PRICE, AND SERVICE

In the Telecommunications Industry, purchasing decisions come down to an evaluation of: **Product**, **Price** and **Service**. Because of this each should be addressed up front. In the first appointment I suggest that you cover them in any order as long as you cover **Price** last.

SALES MANAGER'S EXERCISE:

Role-playing: Either review yourself and customize the following scripts for your first appointment, or have your salespeople review and customize. Have a salesperson play the part of the prospect and role-play the scripts in the front of the room. Have each salesperson prepare:

1. A powerful service and support story using techniques that follow.

2. A powerful, compelling company story.

Practice the stories in front of the group.

SCENARIOS

The first scenario is for those of you selling telecommunication services such as local or long distance services. If you are still selling price only — you're in trouble — because you're certainly going to be able to initially undercut a full-service provider and save your customers some money, but if they jumped to you they're also wide open to be approached by one of your competitors. All they have to do is hear the words, "additional savings."

Early in the first appointment, try to position yourself as the vendor that can provide all kinds of value-added services, such as, consultative services or personalized services or additional product offerings or point to point, as well as, long distance and/or local service.

You are trying to convince the prospect of the benefits of doing business with you and your company. It is up to your sales manager to articulate to you what makes your company different from the competition.

A little preparation goes a long way.

Let's say you are meeting with a prospect for long distance or local services for the first time. You've introduced yourself, and had a couple of minutes of "chit chat," or "schmoozing." as a friend of mine from New York might say, and you've done your best to put the prospect at ease.

Personally, I prefer to use a little "wit." I try to talk about the present situation, about the environment, about what I think the prospect might want to talk about. I usually listen closely to what the prospect says, when I say:

"Very nice to meet you. How are you today?"

Usually, I'll pick up on what they want to talk about. It could be about the weather, about a local sports team, about the circumstances of my being there, about their business, or anything else. One well-known sales trainer cautions young salespeople to "never compliment the fish, hanging on the wall". He argues that every other salesperson compliments the fish.

Never compliment the fish.

So, there you are, the prospect is relaxed and looks you right in the eyes, and says:

"Okay what have you got?"

I can promise you, you better be prepared and not ramble around...or you'll lose out in the very first visit.

Note: For those of you where a great deal, if not all, of the selling takes place over the telephone, this scenario is easily adapted for the first telephone conversation.

Proven-effective First Appointment Script for Selling Local or Long-distance Services

"Well our company has found that telecommunication decisions revolve around THREE MAJOR AREAS OF CONCERN.

"The first major area of concern is SERVICE AND SUPPORT. Maybe there are some items that you can buy for your office...and if the vendor goes away...you'll still be ok...for instance your paper supplier...or temporary personnel agency. It might be relatively painless to find additional sources of paper, or another personnel agency. Telecommunications on the other hand is a lot more complicated. You can survive for awhile without paper, or find replacement paper, or scramble around for a new personnel agency. But if your telecommunication links go down — your company's ability to communicate with the world — your business comes to a halt. This is serious business. We're working with you all the time. So the first thing we must do together is to make sure that you feel totally comfortable with my company's service and support."

At this point, if the prospect agrees, you are ready to establish your credibility as a vendor. It is time to tell a powerful, compelling company story. Think about the things that make your company different, those things that differentiate your company in the marketplace. Here are some suggestions.

- number of years in business
- number of accounts
- type of customers
- major accounts doing business with you
- customer service orientation
- geographic presence

Think for a minute about those aspects of your company that differentiates it in the marketplace. Positive differentiation, from a marketing standpoint is good. If the public, for whatever reason, think about your business or product differently — that is, more favorably than any other company — then they will beat down the door to do business with you.

Take a moment to review the major points that you would make about your company's service and support. Take, for example, the number of years in business, or the number of accounts doing business with you, or the major, well-known accounts that do business with you, or your customer service orientation, such as toll free numbers, or diligent follow-up and support and focus on those differentiating factors. That way you separate your company from all others in the marketplace.

Decisions about telecommunications revolve around three areas of concern:

1. Comfort with service and support.

2. Ability to customize a system to fit your needs.

3. Price.

Differentiate yourself: What I am going to share with you now will be worth all the big bucks you've paid to purchase this book. One of the best ways to position your company ahead of its competitors in the eyes of your prospect is to differentiate yourself. When you talk about service and support, the best, most skilled, and effective salespeople personalize their stories about service and support. In other words, they customize the stories to apply to the situation, and when they find a short, possibly humorous story that demonstrates their company's commitment to the marketplace (and the story produces results), they tell it over and over again.

Tell a Powerful Service and Support Story: The best stories come from personal experienced. Like when you had to call a technician on a snowy Sunday in January, and he/she got stuck in their driveway, and you had to go dig him/her out (plus you had to arrange for a babysitter for their kids) so they could do a repair to a client's phone system in time for a big Monday morning's calling volume. All this time the client was going crazy, not sure if you could find someone over the weekend, but you did, and they fixed the problem, and the client was able to sleep soundly Sunday night.

Formula for telling a great story:

1. The story needs the element of veracity (truth).

2. The story has to have conflict.

3. The story should end with a successful resolution.

You can remember how to tell a powerful story by remembering that the three elements of a good story — veracity, conflict, and resolution — are like a VCR running through the prospect's head.

Let me share with you another example of a service-oriented story. Rather than just telling the prospect how great your company's service is, ask them if you can share a short story that makes you feel great about working for your company.

Learn how to tell a powerful service and support story.

Mr. Prospect, remember last year when we had the terrible floods in downtown and over 300 business were out of service? *(Veracity...true story)* Our company had service contracts with over 75 of those businesses, and you can imagine the turmoil. *(Conflict)* We were able to call in extra technicians from all over the state, and a few even flew in from Chicago. We had every business up and running within two days of the time the water receded. *(Successful resolution)*

Allow the prospect to ask questions and don't move on the your next point until you place closure on service and support.

Now you can move on to the next area of concern.

Product: "The next major area of concern that decision-makers like yourself have is with regard to product...maybe in some industries...the products are similar...and price might be the differentiating factor. This is not so with regard to telecommunications. In our business the SERVICE AND PRODUCT OFFERINGS ARE COMPLEX. The key to purchasing the appropriate product offerings is my ability to understand your company, the industry, how you compete within your industry and to recommend the best offering to fit those needs,"

At this point you can give an example of one of your product or service offerings, but remember, you are not offering any solutions at this time, you are simply setting the stage, differentiating yourself, building credibility.

"The key to purchasing telecommunications services is a very custom project...your telephone traffic and your telecommunications usage is as personal to your company as your fingerprints are to you...and most important your company exists in a dynamic environment, not static, it's always changing."

Now you can move on to their final area of concern.

Price: "If you're like many of the other companies I work with then your third major area of concern will be price. Many times we can effect considerable savings. I won't know until I really analyze your bills and understand your business better, but I can promise you this (here's a suggestion, and it's just a suggestion, to take the pressure off price)...we probably won't be the least expensive vendor you can choose, on the other hand, we won't be the most expensive. We will be price competitive in our offerings against the major competitors, by that I mean 10-15% — don't you feel that is competitive?"

What you're trying to accomplish with the above statement about price is to divert the decision from price alone to all the other value-added services that you offer.

Trial Close: At this point in the meeting some of you will feel very comfortable with your prospect. Maybe you've established a pretty good rapport. Remember you don't really know a lot about their business yet or really about how much you can help them. You're still qualifying. I'd like to share with you now a great qualifying technique, especially for those of you that might have a little trouble closing. This technique sort of opens up the close on the first appointment, and as you've probably guessed...it's a trial close.

You've talked about service, you've talked a little about product offerings without giving solutions, and you've covered price considerations. If you're comfortable, then say this:

"May I ask you a very direct question?"

"If I can make you feel totally comfortable with my company's service and support, and if you had confidence in my ability to fully understand your needs, and can recommend the appropriate telecommunications services for your business, and if I could do it at a competitive price...is there any reason you wouldn't buy it?"

If the answer is positive then you have one more important question to ask:

"Good...but by the way, of the three things we talked about...my company's strong reputation for service and support, my ability to fully understand your requirements and recommend the right system for you, or price, which is most critical to you? Could you rank them in order of importance?"

The prospect has just told you how to sell them.

If the answer is negative, then you're on your way to continued qualification. Maybe the answer is negative because they're not ready to make a decision, or they love their existing vendor, or there are political reasons for not changing, or someone else makes the decision, or they want to evaluate 15 companies before making a decision. In any event, your trial close begins a powerful qualification process.

What you will soon be learning, if you haven't already, is that your ability to differentiate yourself, build credibility, and qualify the prospect in the initial meeting will be critical to your sales success.

If the answer to your trial close is positive, it's possible to move to the next step in the sales process: the fact-finding stage, which includes rigorous qualification.

Proven-effective First Appointment Script for Selling Equipment and Hardware

Now let's look at more appropriate words and script, for those of you selling equipment and hardware.

Remember these are my words. If you don't like them, then you can change them to meet your style, but try to capture the intent of what you're trying to do in this first appointment. Try to keep your objectives in mind:

Objectives for your first appointment:

- Establish partnership type of rapport.

- Find commonalties.

- Set up time frame.

- Qualify.

- Gain trust and confidence.

- Position yourself for the long term.

- Establish your credibility.

- Establish credibility in your company, products, and services.

- Be so prepared that you're in control.

- Gain commitment for the next step

"Basically Mr. Prospect when companies like yours evaluate new telecommunication systems, they have three major areas of concern. The first, of course, is service and support. There might be other products you can purchase such as your computers or printers...and if they fail...that's a big aggravation...and it's nice if the repairman shows up on a timely basis...and if they don't it's a problem...but at least your company can continue to operate. With regard to your telephone system, if that fails...and you don't get prompt service...then your business comes to a grinding halt."

(Now tell your powerful company and personal service and support story.)

"The next major area of concern is the hardware...the success or failure of your installation will depend upon my ability to fully understand your needs and customize a system to fit those needs...taking into consideration present and future growth."

Here you relate a powerful, confident story of how you understood some subtle differences in what a customer thought they needed and what they actually needed and how your approach was different. Here's a simple example:

Recently, I wanted to purchase some scatter rugs to compliment my blue and white furniture in my home on the ocean here in beautiful Manchester-by-the-Sea, Massachusetts. As I was meandering down the isle at the local Home Depot, looking over the merchandise, a salesperson wandered over, and asked if she could be of assistance. When I kind of dismissed her inquiry with some kind of a comment about needing some simple scatter rugs, she asked me how I was going to use the rugs. I have to tell you, that the only thought in my head at this point was finding the right color. I mean how difficult is it to buy a scatter rug. The saleslady was initially concerned that they were the non-skid type. (Wow...I didn't even know there were two different types of scatter rugs.) When I explained that they were to go over my existing wall-to-wall carpeting, she pointed out the importance of getting the kind with a color that wouldn't run or ruin my existing rugs. Now I was a little more knowledgeable about "scatter rugs." I was looking for a non-skid, washable, blue and white scatter rug that won't ruin my wall to wall carpeting if it got wet.

It was the salesperson that pointed out to me that in addition to the color, I needed a non-skid, washable, scatter rug that was color safe (i.e., wouldn't run when wet). In this instance, the salesperson differentiated herself, not the product.

After differentiating yourself with a short story, you would summarize as follows:

"You see, your needs are different from anyone else. Of all the companies I've worked with, no two are the same in terms of configuration, design, growth, usage, capabilities...and they're all installed in dynamic environments."

SALES MANAGER'S EXERCISE:
Have your salespeople think of an example where what the customer actually needed and what they thought the customer needed were different. Role-play the stories.

"Last is price...almost always someone will ask me about price." (Again it's important to take the issue of price out of the formula.) Say, "we will not be the lowest cost vendor that you might consider...but on the other hand we probably won't be the most expensive...we will probably be priced within 10-15% of the major manufacturers...isn't that competitive?"

Here's the key to making sure that you say this up front during the sales process. When it's time to close...and if the prospect tells you that someone else is 10% lower...you can refer back to your initial conversation and remind the prospect that you only promised to be competitive, not the lowest priced.

Next, as in our previous script, try a trial close.

RE-CAP

It's time to review where you are at this point in the first appointment.

With your script and knowledge of where you were going in this first meeting you are well enough prepared to be in control at all times — you've been able to establish credibility — in yourself, and in your company

You were well prepared, knew what you were going to say. However, you didn't appear too polished, although you made a good first impression because you were professional and got down to business quickly.

You began the establishment of a partnership-type of rapport, not a personal rapport. Remember that the prospect is not looking for a friend, just a business associate that can be trusted.

With regard to searching for commonalties, as you talked through your script of service, product and price, you searched for commonalties between your company and the prospect's company. Do both businesses have a passion for service, or a commitment to excellence, or quality? Are they a small, local company focused on the same city and state, or are they a nationwide, multi-site organization? Magnify the commonalties in your conversation. Just the way individuals like to buy from people like themselves, companies like to buy from companies like them.

Try hard to position yourself for the long haul, in that you and your company are committed to this marketplace for the long term.

All of the above should be conducted in a relaxed, friendly, and easy going environment with everything possible being done to put the prospect at ease so you can build credibility and establish a rapport. You have now reached a critical point in the sales process. If you sense there is time, and your prospect is in the right frame of mind, then you can move right into the qualification process. If you sense time pressure, then I suggest you get an agreement from the prospect to allow you to return for another appointment to do a better job of qualifying. (The next step in the sales process.)

What I'm sure you're aware of is that because of the wide variety of products and services in our industry, and the wide variety of costs, the length of the sales cycle varies greatly. The sales cycle can vary from one-call closes — where all steps are completed in one visit or call — to up to a one to two year sales cycle.

We will cover the qualification step in detail in the next chapter, but simply to cover all bases, especially for those of you working on the shorter sales cycles, and consider qualification in the first appointment critical to your success, then before leaving your first appointment you must qualify for:

- The decision-maker (Is there anyone other than yourself that will make the final decision?).

- Stage in the process (If you were to make a decision today can you share with me who would be leading the pack?).

- Time frame or impending event (Could you share with me the decision making process and your sense of urgency?).

- Other competitors (Who will you be looking at?).

- Their budget (What is the budget for?).

If, after a pretty rigorous qualifying session, you still believe there is an opportunity for you, then get a commitment from the prospect to move to the next step — the fact-finding phase.

MY OBJECTIVES FOR A GREAT FIRST APPOINTMENT

Notes

Step 3: Qualifying

If you are presently earning a living, or plan on earning your living selling telecommunications equipment or services then it is essential that you learn the art of qualifying. This chapter is devoted exclusively to helping you develop that skill.

You probably don't need me to remind you that as a salesperson one of your goals is to maximize your time in front of "qualified prospects." You are constantly on the lookout for those prospects that are ready, willing, and able to purchase your products and services. Many of the most successful salespeople in the Telecommunications Industry will tell you that much of their success depends not only upon their ability to qualify prospects, but also on their ability to qualify prospects early in the sales process. One highly successful sales manager, Noel Norwood, now the VP of Training at Gillette Global Network, Inc, and previously of Staples Communications and Wiltel prefers not to call it "the art of qualifying" — he calls it the "art of disqualifying."

Noel and I both know how hard this aspect of the sales process is for most salespeople.

Think of qualify as the "art of disqualifying".

REVIEW

Before we can talk about the qualification process, however, it is important to remind you, once again, that as we work our way through the sales process, each Chapter builds upon the previous chapters.

The primary objective of Chapter 1 was to ensure that you feel great about the profession of selling in general and are truly excited about being in the Telecommunications Industry.

Chapter 2 laid out an overview of the full eight-step sales process from Prospecting to Closing to Follow-up and Support.

In Chapter 3 I explained and showed you the importance of prospecting and how to perfect your prospecting skills — Step 1 in the selling process.

Then in Chapter 4 we discussed exactly what to say, do, and how to act as a salesperson on your first appointment or during your first telephone conversation. We also pointed out the importance of developing the skills to enable you to forage for the right prospects.

LET'S TALK ABOUT QUALIFYING

Some of you might be saying to yourself at this point:

"That's okay because my company finds the prospects for me...I just go out there and close."

It's been my experience that salespeople who just meet with so called "qualified leads," and closing prospects usually are not the ones making the big bucks in this or any other industry. Anyone can find order takers. In our industry the really big bucks are reserved for salespeople skilled at finding prospects, as well as, meeting with prospects, making a great first impression, then qualifying prospects, making great presentations, selling value, establishing wonderful relationships, and closing sales opportunities.

So once again, I need to stress that before you can begin this step in the sales process — the art of qualifying — it is crucial that you have perfected the previous step — the art of the first appointment. You might want to take a minute to quickly review some of the lessons learned in the last chapter.

Make sure at this point, that you and your sales manager have a full understanding of exactly what should be accomplished in that first appointment.

Many sales managers, depending on the products you are selling, will prefer a full qualification process to occur in the first appointment or conversation. Because of the importance of qualifying and because qualifying — really qualifying — can put some pressure on the sales process, I decided to devote an entire chapter to this part of the process. You should decide now with your sales manager the timing of the qualification process. I don't really care as long as it is done thoroughly and early (relatively speaking) in the sales cycle.

So now let's talk about qualifying or questioning. If you remember, at or near the end of our first appointment or contact with the prospect, we used a trial close or questions to assist in starting to get the prospect involved in the selling process. It is truly impossible to help a customer buy, without first asking questions.

Harvey Mckay in his popular book entitled "Swim with the Sharks" says:

"Knowing something about your customer is just as important as knowing everything about your product."

SALES TIP:
- Notice what your customer likes
- Ask what your customer likes
- Learn what your customer likes
- Use the information to bond

When we talk about qualifying, aren't we really talking about building relationships? Indeed, in selling, as in all of life, certain guidelines apply. If you truly want to build a lasting relationship, then it is important to notice what your friends, relatives, loved ones, customers like, ask them what they like, and use the information to bond and build relationships.

With regard to questioning and qualifying let's explore ideas that have helped me immensely in my selling career.

CONCEPTUAL SELLING

First of all, let's talk about the process of conceptual selling. The idea here is that you still must look at selling as a needs/benefit game. In other words, you are still asking questions to uncover needs or problems. With regard to telecommunications, the needs or problems are usually quite straightforward. They can be as simple as "we're paying too much per minute for a long distance telephone call." Or the prospect may only need solution to a problem such as the "customers are complaining about being put on hold too long during or prior to the order taking process."

Certainly, you can see that as a sales rep, there are pertinent questions that must be asked. Unless you ask "how much per minute are you currently paying for long distance calls," or "do customers ever complain about being put on hold or about extraordinary delays during the order process," you will never know there is a problem, or that the issue is of concern to them. Questioning about a customer's potential problems is an essential part of the sales cycle, but in conceptual selling you ask questions that encourage the prospect themselves to discover that they have a problem.

In conceptual selling you still ask questions to find a need, or explore a capability. But instead of you, the salesperson, relating their need and then relating the benefit of your products and services as the solution, you simply set the stage so the prospects will find the need for themselves.

I want you to think of conceptual selling as a skill that creates an environment where the bridge from need to benefit is self-discovered.

To become great at conceptual selling you must know your products' capabilities perfectly; because in this kind of selling you will be leading the prospect to the conclusion that your solution is the best, so when you get there, you better have the right solution.

> **SALES MANAGER'S EXERCISE:**
> To demonstrate the importance to your salespeople of asking great questions follow the example below, practice a few times and do it yourself at your next sales meeting.

The ESP Exercise

Let me give you an example of a little showmanship that I learned from Tom Hopkins, one of the most well known sales trainers in the country. Here's the idea. When I'm speaking, I'll pick a person out of the audience and ask them if they have special ESP (Extra Sensory Perception) skills. Of course they say no. Then I say, "we'll see about that." I state that I will ask that person to pick a card at random out of a pack of cards, and I'll look at it, and the audience will look at it, but that person won't. Through the process of that person asking some simple questions, he or she should be able to guess what card was chosen out of the deck.

First, they pick a card, I'm careful that they don't see it. I look at it, the audience looks at it, and we put the card aside from the rest of the deck. Let's say the card is the Queen of Hearts. Here's how I begin: I start with easy questions, questions they can answer easily, just like the way you would start in your questioning techniques.

"Are you familiar with a deck of cards?"

"Yes"

"Are you aware that there are face and number cards?"

"Yes"

"Of the two types of cards, face and number, which do you prefer?"

Now here's the key to this little exercise; if they say, "Face", you say, "Great...I like face cards too." (Remember, you are leading them toward the Queen of Hearts.)

If they say, "The number cards", I say, "Great, so that leaves the face cards...right?"

"Now within the face cards (divide them in half)...there are the Jack and Queen, and the King and Ace...of those two combinations which do you prefer?"

If they say, "Jack and Queen," I say, "Great so do I." If they say, "King and Ace," I say, "Good...that leaves the Jack and Queen," and keep right on going. "Now, between the Jack and Queen which do you prefer?"

If they say, "Queen," I say, "Great...so do I." If they say, "King," I say, "Good, that leaves the Queen. Is that correct?"

If they say, "Right, the Queen," then I say, "Good...now, are you familiar with the suits in a card deck?"

They'll say, "Of course," and without me telling them (just like you not relating the actual benefits of your solution or why your company, or product is so great) they say, "Spades, Clubs, Hearts and Diamonds."

I say, "Good," in an encouraging kind of way. Just like you maybe saying "OH," in the questioning and qualifying stage to elicit more information from a prospect. To have a little fun and usually to get a little chuckle or laugh, I'll say, "Now let's not make this too confusing...are some red and some black?"

"Yes."

I could have done the differentiating for them, by saying: Spades and Clubs are black, and Hearts and Diamonds are red. You see this is just like the way you know how your product works. You know the real benefits of your technical solutions. The concept in conceptual selling is for you not to tell the prospect the benefit, but to ask the question and have the prospect say the benefit. Try to teach yourself to hold back and get the prospect, just like the subject in my exercise, to make the connection.

So when the student/audience member says, "Hearts and Diamonds are red...and Spades and Clubs are black," I say, "Great, which do you prefer red or black?"

If the person says, "Black," I respond with, "Great, that leaves the red suits." You see, if your tactics are sound, they can't fight you.

"Now within the red suits which do you prefer...the Hearts or the Diamonds?"

If the subject says, "Hearts," I then state, "Terrific...I love Hearts...so your card is the Queen of Hearts!"

The audience roars, and the student is mystified, and wondering; "How did he do that?"

I've done this little trick or exercise many times, and am amazed how many people don't get it, or don't really understand how I did it. The purpose is simply to demonstrate the effectiveness during the sales process of asking not telling.

Never Tell Always Ask
Let me share with you three simple examples.

Let's say you represent AT&T as a re-seller and Sprint for certain special applications. First, don't tell your prospect that Sprint is the most cost-effective solution for their particular calling patterns, but ask them if they have a preference of AT&T over Sprint? You can then tailor your recommendation depending upon their response.

Second, we will take it a step further and say your phone system is fully digital, and you're competing against an older analog model. Resist the urge to explain the advantages of digital over analog, but ask the prospect if they are aware of the advantages of digital over analog.

And third, assume that the hardware you represent has the ability to record a telephone conversation, simply by pressing a button during a call. Don't tell the prospect about this capability, but ask; "Do you ever, or do you see the need, to be able to record a telephone conversation during a call."

In the telecommunications industry, almost every capability can be interpreted as either a negative capability or a positive capability. With regard to telephone systems, I've seen this over and over again with such simple features as off-hook voice announcement, camp-on, and call forwarding. Even during the early stages of voice mail, with all of its inherent time saving benefits, there was a tremendous backlash since people hid behind their voice mail. Therefore, by telling rather than asking, many salespeople ended up putting their foot in their mouth. It is better to ask, then listen, and respond accordingly.

Conceptual selling takes great listening skills. First you must never ask a question where you're not prepared for the answer. Second, you have to really listen closely once you ask the question. By the way, the best way to *listen* is to rearrange the letters in the word listen. Ironically, they can spell *silent*.

So why did I take the time to relate to you this technique? It is because this art is the essence and the heart of selling. **Never tell — always ask.** If you tell your prospects things, they'll fight you. If you ask the right question and they answer it, the answer becomes theirs. They will own the solution, and that is really selling. Selling is nothing more than helping people to make decisions that are good for them.

The Indonesia Story

A couple of years ago, I traveled to Indonesia to conduct a sales training class for LM Ericsson Telecom in Jakarta. It was my first trip to Indonesia, and as you can imagine, when I was there I learned a lot about the country. One thing I learned early on was that Indonesia is the fourth most populous country in the world.

Shortly after my return to the states, I made the following statement in a seminar: "Indonesia is the fourth most populous country in the world." Wow, did I get an argument. "Couldn't be...what about the U.S., and Russia, and India, and China, France, Italy, Spain..."I don't even know where Indonesia is?" replied one student. If only I'd asked the question: Can anyone guess what the fourth most populous country in the world is? Someone might have guessed Indonesia, and everyone would have been happy. I would have been able to lead them to the conclusion by maybe first asking who knew where it was and how many islands (13,000) were in this huge island mass.

Uniqueness Sells

So, with regard to conceptual selling or selling by questioning, why is it so great, so powerful? It is used by less than 1% of salespeople, thus it is unique, and uniqueness sells.

Conceptual selling thrives on the deepest of human characteristics, that of self-discovery and intellectual excitement. Sales and sales training techniques change and evolve over the years, but certain things will never change. Such as human nature will never change. It is not a pushy technique, it is in fact, "pully." Conceptual selling demands the use of tons of emotion and empathy for the prospect. It is a powerful selling technique that is less of a science, and more of an art. You can use it to its fullest or only use it occasionally during the sales process. No matter how much you use it, it is very powerful. It requires self-control, the ability to hold back, and contain yourself. The best part of the concept is that if you mess up, you have lost nothing, you simply revert back to the old-fashion, yet still effective method of needs/benefit selling.

Conceptual selling gives you the pleasure of watching self-discovery.

It requires total honesty and honesty sells.

It demands excellent product knowledge and I can't stress enough the importance of developing excellent product knowledge.

It demands excellent competitive knowledge.

You see, by using conceptual selling, you can lead your prospect toward their own conclusions. If done right you lead them to a conclusion that they need what only you can offer not what you and your competitors offer.

The Pen/Pencil Exercise

Many times in my seminars, I'll ask the audience to take out a pencil or pen. Then I'll tell them to turn to their neighbor and sell them the pen or pencil. Often, I'll catch even a veteran salesperson holding up a pen and start right in:

(Let's say it's one of those expensive Cross gold pens)

"This is one of the finest pens on the market...it comes with a replaceable cartridge and opens easily for changing...let me demonstrate...see how beautiful the ink looks on the paper and it comes in blue, black, and red.... This is the kind of pen that you'll want to hand down from generation to generation...and so forth and so forth...and so forth...blah...blah...blah."

This is all fine...but what if you're selling to someone who loses pens all the time. You're telling them how great this pen is and they're thinking how often they lose great pens like that.

Or I'll catch a salesperson, holding up an inexpensive, 19 cent BIC pen, and saying:

"This is the most inexpensive pen on the market today...this is the kind of pen...if you're the type that loses pens all the time, you can lose one of these a day and still save money over buying those expensive pens...and what's nice is you can see exactly how much ink you have left...and when you run out of ink, you can just throw it away."

(Yeah, sure. Is that before or after, the pen explodes in my shirt pocket and I get ink all over my $40 dollar shirt, or ruin a pair of pants.)

The salesperson is thinking that an inexpensive pen is cost-effective, and you, the prospect, are thinking about how many times an inexpensive pen has exploded in their shirt or pants pocket.

SALES MANAGER'S EXERCISE:

At a sales meeting, or when interviewing a prospective salesperson, take out a pen, and ask them to "sell me this pen." You will learn instantly how professional they are, how well they have been trained, or whether you have a project on your hands. Judge how much they talk versus how many questions they ask.

With regard to the pencil/pen example:

Real pros introduce themselves, then ask how their neighbor is doing. Such as, "Are you enjoying the seminar? Good, so am I." They also do a little bonding, and then they gently ask: "Do you mind if I ask you a couple of questions...do you write? Do you prefer pen or pencil?"

If they answer "pen" "Do you prefer expensive or inexpensive pens"...and so forth.

SALES TIP:
Never try to sell anything to anybody without first finding out what's important to them. When you ask the right qualifying questions, they'll tell you exactly how to sell them.

PRESCRIPTION WITHOUT DIAGNOSE IS MALPRACTICE

Our job, as telecommunications salespeople, at this stage of the sales cycle is to uncover needs. Give no solutions at this point, only after you find out about all of the prospect's needs do you then match those needs with your solutions. Don't talk about features, but do talk about benefits.

Here's a wonderful quote that I first heard from David Finch, the VP of sales and marketing at ATCOM in Charlotte, NC. When fact-findings and qualifying you must remember that "prescription, without diagnosis, is malpractice."

SALES TIP:
When fact-finding and qualifying you need to remember that "prescription, without diagnosis, is malpractice."

When questioning you are diagnosing, probing, analyzing. Then and only then — after you have the facts — can you prescribe the correct solution the way a doctor prescribes the right medicine.

From time to time, I've had a salesperson ask me in one of my seminars to advise them what to say, when a prospect asks them early in the sales cycle; "How much is this going to cost?"

I suggest something along these lines:

"Mr. Prospect, I'm not sure at this *stethoscope on chest* point exactly how much this is going to cost, because at this moment I don't really know what you might need. Have you ever heard the expression: "prescription...without diagnosis...is malpractice.""

In sales, like medicine, "prescription without diagnosis is malpractice."

When you become a master at questioning you'll become a great salesperson. Let me share with you a vivid example.

THE PENNY CANDY STORY

My 21 year old daughter, Debbie is smart, and as of this writing is going into her Senior year at the University of Pennsylvania. For as long as I can remember I've taught Debbie never to whine. My mother and father hated whining and I seldom whined, plus it never worked anyway. They just said on my occasional moments of weakness, "stop your whining," and that was the end of it. Debbie never whines, but she has managed to get just about everything she ever wanted.

You see, she's not only the daughter of a professional salesperson, but the daughter of a professional sales trainer, and over the years I guess she's picked up a lot more than I thought. You've probably heard the expression: "The apple never falls very far from the tree."

The apple never falls far from the tree.

During the years when she was attending Shore Country Day School in Beverly, Massachusetts, I'd take her and pick her up at the bus stop in the middle of our town. There was a great penny candy store nearby called "Gilly's" that was owned by a good friend and running buddy of mine, Jim Gilford. Debbie, in her early years, after she got off the bus and into my car for the short ride home, would ask:

"Hey Dad...can we go to Gilly's and get some penny candy? Inevitably, I'd respond, "Sorry Deb, I've got to get right back to my office...and make some more phone calls...or mail out something before the post office closes at 5:00 or I'm expecting an important call, or something to that effect.

Debbie was in complete control of the "sales" process.

Soon she learned the technique we're talking about in this chapter. She'd jump in the car, all smiles, and ask me how my day's going. When I'd finish responding, and I'd ask her how her day went she'd excitedly relate a great story, usually pulling out some unbelievable piece of art work, or school project (all the time using great restraint and holding back her real intention). Then just at the right time (timing is everything), she would say something like:

"Hey dad...remember back when you were 7 years old?"

"Yessssss."

"Didn't you every once in a while have this incredible urge for something?"

"You mean like penny candy at Gilly's?" I'd blurt out.

"Yeah...well I'm having one of those urges right now. ...And Dad...think how much fun you'll have pretending you're a 7 year old again...and guess who you'll get a chance to say hello to?"

"I know, my good running buddy...Gilly...In fact, I do need to ask him about the starting time of the race this weekend. OK, let's go."

It was a closed deal.

So what happened here?

She established a good rapport. "Hey dad, how's your day going?"

She used good questioning technique. "Hey dad do you remember back when you were 7 years old?"

She used lots of emotion. I have nice memories about my childhood.

She used good restraint. She never said, "Penny candy at Gilly's." I did.

Her comments were benefit oriented. I'll have fun, and get to see my friend and running buddy.

She got what she wanted, with no whining.

Did she tell me I needed a work break, and that I could see my good friend Gilly? No.

Did she lead me to the conclusion that I needed a work break, and a chance to see my good friend Gilly? Yes.

Who was in control? Are you kidding?

GREAT QUESTIONING TECHNIQUES

There are four reasons for developing great questioning techniques. First of all, and this is pretty obvious, it is how you uncover the telecommunications requirements and needs and problems. It is also how you qualify your prospects and with the proper questioning techniques you can build rapport and endear yourself to your prospects. And, when you have developed the skills necessary for asking the necessary, right and great questions, you are always in control.

> **SALES TIP:**
> **Four reasons to ask great questions:**
>
> 1. You determine the prospects needs and problems
> 2. You qualify
> 3. You endear yourself to the prospect
> 3. You're in control

Funneling Technique

With regard to the first reason for asking great questions, I've always felt that the funneling technique works best. Start with the most general, easiest to answer questions, then get more and more specific.

I start with questions about their industry.

Could you tell me a little about your industry and your position within the industry?

Are you an industry leader, market share leader, or new entrant?

Is the industry growing, expanding or contracting, changing?

How is it changing?

Of course, some prospects might be tired of having to explain all of this over and over again, and might even feel like you haven't earned the right to ask these kinds of questions.

Think for a minute how much more effective you'd be if you'd done a little homework before making this sales call. In other words, doing some homework on the company so you could ask a question like this:

Learn how to ask great questions.

"Before we get to your telecommunications needs could you help me understand a little better what's going on in the Jet Ski Industry? I read last week in

Business Week that while the growth is phenomenal, many of the manufacturers like you are under a great deal of pressure from environmentalists, and skyrocketing insurance expenses related to liability?"

After the answer, you could continue to probe:

"So even in the light of these financial pressures, will you still be able to maintain your dominant position as market leader?"

"Does that mean you'll continue to sign up distributors, and expand beyond your three major manufacturing plants?"

Can you see how effective having a little information is rather than simply asking: "Do you have any plans to open more manufacturing space?"

Start with the industry, then ask questions about the business, and then about their company, then about them within the company, then move to telecommunications needs.

> **SALES TIP:**
> **When questioning use the "funnel technique."**
>
> First ask questions about their industry then ask questions about their business and lastly ask questions about their telecom needs.

Remember the importance of positioning yourself for the long term. Be interested not just in their company today, but in the future.

How long do you think this rapid growth in your industry will continue; and what could cause it to stall?

During the questioning you must focus, not just on how your products and services can help them solve their problems, but on ways your products and services will help this prospect be more competitive in their industry and serve their customers better.

You could ask:

Could you share with me a little information about your customers? For example, how do you distribute your products, or how do you maintain a competitive edge? Why do your customers buy from you? Do they buy directly, or through distributors? What are the primary methods you use to promote and market your products? What marketing strategies do you use?

Here is a sales tip that was mentioned to me by Michael Weatherly of the Telecom Group in Troy, Michigan. Michael has had a great deal of success qualifying in the industry. During the qualifying or fact-finding stage, he'll gather all the information he needs, but he will purposely leave out one or two critical questions. He says this gives him a reason to call the prospect back, follow-up, bond a little more, and the prospect usually respects him for his attention to detail. For those of you that know Michael, you're probably saying to yourself, "Sounds like something Michael would do." Michael's customers love him.

With regard to asking questions about their company, you have to use your own judgment. There is no end to the level of detail that you can go into. Think about these questions:

- What is your sales and marketing philosophy?

- How are they positioned in the marketplace?

- Do you conduct training and what kinds?

- What is the company's level of technical support, for what products, and in what detail?

- What is the level of expertise of these people?

- How are they trained?

- What is your product niche, and how are you viewed within this niche?

- What is the company's public image?

- How do you distribute your products? What channels?

- What are the various departments within the company, and the number of people in each?

I decided to focus of this book specifically on selling in the Telecommunications Industry because among other things, my strong belief in the uniqueness of the industry. Selling in telecom requires a number of skills that differ from other industries. In telecom our products and services many times impact the entire company and everyone in the company and the decision-makers that you are talking to know this. They know that the decisions they make with regard to your products and services are crucial decisions, and can directly impact their company's success. Those decision-makers will expect you to understand their entire business. Your success in sales will depend on the degree to which you have demonstrated an understanding of your prospect's business. Great salespeople are usually great questioners.

When questioning, first ask general questions about their industry, then more specific questions about their business, then focused questions about their telecom needs.

Discovering Their Telecom Needs

You've asked great questions about their industry and their business or organization now you need to move toward more specific questions regarding their telecom needs. By knowing the right questions to ask you'll be able to determine which of your company's products and services the customer would be able to benefit from, and why. This allows you to make the proper recommendations, support those recommendations, solve your prospect's problems and make sales.

Divide the telecom needs into three areas - incoming calls, outgoing calls and internal calls — with each area cover as many of the areas of communications as appropriate: such as, voice, data, image, fax, cellular, private line, interstate, intrastate, international, calling cards, etc.

Try to not skip around, instead Say something like, "First let's talk about the nature of your incoming calls?" You can talk about the operator, after hours calling, order taking, incoming support, 800 #'s, employees calling in, service technicians calling in, and much, much, more.

Within this process, you also begin qualifying your prospect and start the fact-finding stage. Fact-finding is so important that I'm devoting the entire next chapter to it.

Qualifying

Let's now move to the second reason for learning to ask great questions — to qualify.

How do you as a salesperson know if you've fully qualified a prospect? By knowing that there are a number of variables that need qualifying. But, how do you remember to qualify for everything? During my work for Claricom, a nationwide telecom distributor, and originally formed from the direct distribution arm of Executone, I became familiar with Executone's qualification methodology. The method of remembering the variables to qualify, comes from Executone's well known Keystone training, which was designed specifically for salespeople selling equipment, the salespeople are taught to memorize the acronym: MANIAC T

It stand for:

Money — Have they got the money, or can they get the money for your product? Have they budgeted any money for the project? Do they have an amount in mind for the project?

Authority — Are you talking to the person that can write the check, or get the financing? Do they have both the authority and the influence to get your recommendation approved in their organization? Are they the decision-maker?

Need — Why do they need the new stuff? What is their problem, and how will it impact their business? What would happen if they don't go ahead?

Impending event — What will cause them to have to make a decision? Is the event clearly defined, such as a move or expansion, or a merger?

Application — What are they going to do with your stuff? Are they interested in Voice Mail, Automatic Attendant, Call Detail Recording (CDR),

Interactive Voice Response (IVR), Computer Telephony Integration (CTI) or Automatic Call Distribution (ACD) type applications?

Competition — Who else will they be looking at, and why? Who is the incumbent, who is favored? If they were to make a decision today, who would be leading?

Time Frame — When will they make a decision? What has to happen, or what events must occur between now and then?

> **SALES TIP:**
> To easily remember what to qualify for, remember the acronym "MANIAC T" which stands for:
> **M**oney
> **A**uthority
> **N**eed
> **I**mpending event
> **A**pplication
> **C**ompetition
> **T**ime frame

The above acronym covers the basics and you'll be well on your way to becoming a great qualifier if you cover the above listed issues. You could also qualify for your prospect's likes and dislikes, by asking:

If you had a wish list, what would you like to see in your telecommunications provider?

If you were to invest in a new telephone system today, what would you alter, change, or modify, or add to your existing system?

You could qualify about their decision making process with the question: How will you go about making a decision of this magnitude?

When you are asking specific qualifying questions, here are a few things to think about. Think of this qualification stage in the sales cycle as a two-way street. Let the prospect qualify you as well. Validate that they feel comfortable that you can be their telecommunications provider.

Be careful of the intensity of your qualifying questions. If you qualify to heavily, too early, you will alienate the prospect. On the other hand, fail to qualify fully, and you could waste a lot of time, or spend too much time talking to the wrong person.

Build Rapport

The third reason for learning how to ask great questions is that you build rapport. People love to talk about themselves, and right after that, they like to talk about their company. Especially if they own or run the company. It is just natural for them to like someone that shows interest in them or their company. Think about the last party you went to. Do you recall a person you met that made a positive impact upon you? I would bet that person made you feel important because they listened to what you had to say and encouraged you to talk about yourself, or your work.

If you're not convinced of the huge impact that showing interest in someone can have, then try this at your next family picnic, or gathering. Go to an aunt, an uncle, a grandmother, or grandfather, someone at least over the age of 65. Innocently enough during the course of the conversation ask this question. "Hey Uncle Charlie, or Aunt Sara, good to see you again, by the way how have you been feeling lately?" Then shut up and listen, nod, grunt, say oooh, aaah, encourage, but don't interrupt. Within a week or two of the event, you'll get positive feedback from your mother or father. "Wow, what did you say to Aunt Sara, she just went on and on how great you were."

The same principle applies to selling. By asking great, well thought out questions and encouraging your prospects to talk about themselves and their business, you're not only gathering valuable information and qualifying, but you're also laying the foundation for a long-lasting relationship. In the next chapter on fact-finding we'll look at 10 additional and critically important guidelines to support your questioning techniques.

Maintain Control

The fourth and last reason for asking terrific questions is it helps you maintain control. Control is critical during the sales process, and it is easy to lose control. In Chapter 6, I'll test you to see how well you stay in control in these early stages of the sales cycle. Then I'll show you how to stay in control, give you objectives to follow, and a fixed strategy to guide you. You'll have the ability to react to questions and guide your prospects down the desired path. Wasn't my daughter

Debbie in control at the bus stop? Wasn't I in control when leading our volunteer to guess the Queen of Hearts?

RE-CAP

So there you have it — qualifying and questioning. Next up is an entire chapter devoted to fact-finding (Step 4). You will learn in Step 4 exactly how to apply your new found skills of questioning and qualifying to the process of fact-finding. In other words, gathering the kind of telecommunications insight that gives you the perfect set up for the next step in the sales process — the Presentation.

Just a reminder that your success depends upon your:

- Attitude

- Skills

- Knowledge

And the first letter of these three spells "**ASK**".

Step 4: Fact-Finding

This chapter is all about fact-finding. The needs analysis — probing and surveying — that allows you to establish what your prospect's telecommunication needs, problems and concerns are all about. In other words, acquiring insight into how your prospective customer communicates with the outside world.

You will learn how to develop a complete understanding of your prospects and their wants and needs so that you can provide intelligent answers to any of their telecommunication concerns.

REVIEW

I'd like to begin by once again reminding you of how this book works. Each chapter builds upon the earlier chapters. Selling is a process, so I want to caution you about reading this chapter on fact-finding unless you have read the preceding chapters.

Let me share with you a short story that illustrates exactly what I mean.

RE-FUELING STORY

As a junior officer on the U.S. Navy Destroyer USS Preston (DD-795) in Vietnam, one of my responsibilities was to supervise the forward re-fueling station. You see, off the coast of Vietnam in the Gulf of Tonkin, it was necessary for our Destroyer to go almost every day alongside other ships for the purpose of refueling, re-arming, and re-supplying.

If you haven't seen this, or are not familiar with this exercise, here's how it works. The aircraft carrier, tanker, ammo ship, or supply ship would steam along at a steady 15 knots, on a steady course. Our destroyer would approach from the stern, and usually come up on the other ship's starboard side, staying about 500 feet away. As we approached, I would supervise the throwing of the bolo, or shooting of the line gun. (A boswain mate would shoot a gun, firing a projectile that was attached to a thin line, usually aiming at a designated amidship location on the other ship.)

Make sure you're well connected before proceeding.

A sailor on the other ship would retrieve the thin, light, line and begin to reel it in by hand. This line was attached to a heavier rope; shipmates would join the lone sailor, they then would continue to pull the heavier rope across to their ship. This rope was then attached to a windlass or crank on their ship, since the rope was connected to an even heavier, thicker cable. If we were refueling, the cable was attached to our refueling hose that was then pulled across and attached to their refueling spicket. Then and only then, when everything was hooked up tight, did pumping of the fuel begin.

I related this story to you, to reinforce two critical aspects of the sales process.

First, I want to caution you about fact-finding too early in the sales cycle.

You can ask probing question if you want, but there is a real danger that at this stage you will get garbage. Think for a minute about the last sale you lost, where you really thought you understood the client's application and telecommunications needs, and really could solve their problems — only to lose to the competition. Did your competitor understand better exactly what the client wanted to accomplish?

I want to make a strong case for imploring you to make sure that you have first established a meaningful rapport with the prospect (Chapter 4). Then, that you have learned how to endear yourself to the prospect and how to qualify by asking great questions (Chapter 5) before launching wholeheartedly into the fact-finding mode.

For those of you that are not sure what I mean by this, let's take a moment to review, exactly where we are in the sales cycle, how we got there, and what we are going to do in this chapter.

I want to once again congratulate each and every one of you for taking time to read this book and also to invest time in yourself to improve yourself professionally. It will do you no good whatsoever to perfect the art of fact-finding, unless you feel really good about who you are, and what you do for a living. If you're not totally comfortable about the profession of selling, and specifically selling in the wonderful Telecommunications Industry, go back and re-read Chapter 1 NOW.

Second, I want to restate that in Chapter 3 we recognized the importance of and then perfected the art of prospecting. Since, why study how to fact-find about prospects, before you're good at first finding prospects to talk to.

In Chapter 4, I discussed exactly what to say in the first conversation, and what to do in the first appointment. I specifically covered how to build credibility in you, your company, products, and services and how to establish rapport early in the sales cycle.

Then in Chapter 5, because of its critical importance in the sales-cycle, I covered the art of questioning and qualifying, and the importance of maximizing your time in front of only "qualified" buyers. Then, and only then, is it appropriate to start the fact-finding stage.

You can see, we're talking about a process here — a thoroughly thought out, extremely effective process — that is well-founded in human nature. A process that, if followed, will produce astounding results. First, build rapport and credibility, then qualify, then, and only then, can you fact-find effectively — first shoot over the thin line, then the thicker rope, then the heavier cable, then and only then the refueling hose.

I want to give credit to a highly successful salesperson, Al Chechatka of WDC Telecard in Boca Raton, for reminding me of the Naval analogy...and the second reason for sharing with you a little of my Naval officer background. Al does all of his selling over the telephone. He knows that if he gives the prospect too much to handle at once he will usually fail. He said that he thinks of the re-fueling analogy to remind him to first build rapport, then to build credibility, before he tries to sell anything. His *modus operende* after the first phone call is that he'll fax the prospect some written and graphic information about his company, then, talk again and do some qualifying, then and only then, begin to fact-find...some things can't be rushed — give your prospects a light, thin line to handle first. Then build rapport and bond; build credibility in you and your company's products and services with some written material — a little heavier line to handle. Then you qualify — the metal cable, if you will. Only after you are well connected, can you begin to really exchange meaningful information — the refueling line.

Now that you completely understand the importance of following the process outlined herein, I want to move to fact-finding. However, it is vital that everyone understands the results of the qualifying stage.

QUALIFYING

What do you think are the three simple questions with regard to qualifying that you should be able to answer before starting the fact-finding stage?

If you are alone, now is a good time to put down this book and recite the three simple questions that you should be able to answer confidently before starting the fact-finding stage.

SALES MANAGER'S EXERCISE:
Ask your salespeople to recite the three simple (but vital questions) they must be able to answer about the prospect upon completion of the qualifying stage. Present some kind of reward to the first salesperson to recite the questions correctly.

The Questions are:

1. Why will they buy? What is it about their existing telecommunications system, or existing telecommunications services provider that is causing a change?

2. When will they buy? Hopefully, this prospect that you're getting ready to spend a lot of time with, has some kind of impending event or driving mechanism that is causing them to have to make a decision, and soon. And hopefully they have budgeted the money for the project, or have sufficient credit, or the financial wherewithal, and you know who will make the recommendation, and who will make the decision.

3. Who will they buy from? One of the first lessons I was taught in sales was to never try to sell anything to anybody, unless they have first told you how to sell them. Sounds interesting, doesn't it? I'll say it again...never sell anything to anybody, unless they have first told you how to sell them.

With regard to the Telecommunications Industry, and within the context of the questioning and qualifying skills previously learned — to have qualified for the third and final important question, you should have asked your prospect:

"Have you ever been involved in the decision making process to acquire a new, major piece of telecommunications equipment?

or

"Were you involved in the decision to choose your existing telecommunication services provider?

Like a prosecuting attorney asking questions of a witness, never ask such a question unless you have a 100% predictability of what the answer will be and then you must know how you will respond.

I can assure you that the answer will be either yes or no!

If the answer is yes, you say "Well great. Can you share with me some of the things that you took into consideration when making that decision?" Again, you want 100% predictability. I can assure you that the answer will revolve around three, and only three critical areas: *Product*, *Price*, and *Service*

The next part of your qualifying conversation is to probe enough to determine the relative importance of each of these to the prospect. When they tell you, they are telling you how to sell them and if they will buy from you — that's qualifying. Those of you that are good at this (I call you qualifying animals) know how to get right at it…build credibility, then qualify and you also know how to disqualify — not waste your time.

But you're a qualifying animal so don't stop there. In today's competitive world, there is always competition. You know enough that if your prospect seems to be holding back a little, for example, during the conversation about vendors, to say: "Do you think I'm number one?" How else will you find out if they're going to buy from you, if you don't ask the question? Or perhaps you can ask, very directly, when the timing is right (assuming that you've done a great job of bonding): "What other vendors will you be evaluating?" and, "Is there anyone that stands out with regard to product, price, and service?"

If the answer is no. Then you have to ask: "What do you think would be some of the things you might look at in a vendor?" And you might need to ask leading questions about product, price, service.

So now that you are sure about the qualifying stage, and can give intelligent answers to the three simple questions:

1. Will they buy?

2. When will they buy?

3. Who will they buy from?

You are in position to take the next step.

EFFECTIVE FACT-FINDING

It is time to move, with full confidence, into the fact-finding stage. The best way to think of this stage is to consider it as gathering information for your presentation, demonstration, or applications-oriented telephone conversation that will follow in the next step.

Each piece of information is like a bullet for your gun.

Obviously the more bullets you have in your gun, the better your chances are of hitting the target. Although I am not "married" to the bullet and gun analogy, I do know that probably many of you have heard it. Selling has nothing to do with shooting or killing: prospects aren't targets. The minute you start thinking this way you will fail. Think of selling as simply "the art of helping people/companies make decisions that are good for them." Obviously, the more information you have about them and their telecommunications needs (and this is where the bullet and gun analogy can be used) — the more bullets you have in your gun, i.e., the better your chance that you and your company will be chosen over the competition to solve the prospect's needs.

Here are twelve suggestions for effective fact-finding:

1. Ask, listen, ask, listen — always hold talking to a minimum, and never interrupt them when they're talking.

2. Ask closed-end, or "yes" questions to which you know the answer. This prompts the prospect to recognize shortcomings with their existing service provider. Then you can follow-up with, for example: "You told me you were concerned about the cost of your technicians calling back into your company from all over the US...Do you agree that spending $.45/minute for the DDD rate is outrageously high?" The prospect's response: "Yes!" Or you might ask: "Don't you agree that having your receptionist write out 30 messages a day at the switchboard is a pretty inefficient way of doing things?" Again the prospect's response: "Yes!"

3. Take lots of notes and listen attentively.

4. The best way to be sure you're listening attentively is to look your prospect in the eyes, smile, nod, encourage, and always assume that what your prospect is saying has value. This isn't always easy, but remember, that's why you've learned how to "qualify", before "fact-finding." Wouldn't you always listen attentively, if you were assured that you were listening to a qualified prospect?

5. Never offer any solutions...bite your tongue.

6. Cover the entire incoming flow of telecommunications traffic to discover exactly how the prospect's company communicates with the outside world. Start with the incoming flow of traffic for voice, data, image, fax and mail. Find out who calls in, when calls occur, where the calls originate — local, national, international, what class of customers call, what's the priority of the various calls, who answers the numerous calls. Don't stop there. You need to establish what happens if a line is busy (who else answers, how do they find them) and/or who takes a message, what happens after hours and on week-

ends, who pays for the incoming calls, what's the order flow and customer service flow. (If you don't have a form, you should design one to prompt you so that you can make sure you've covered all the ground necessary for your particular products and services.)

7. Next you move to the outgoing traffic flow. Who calls out, where and who are the busiest departments, is an outside line always available, do they keep track of the calls, do they allocate the cost, where do they call, what are their traffic patterns (local, national, international), how much are they spending on calls to each area? Obviously, depending upon your products and services, you might have a different set of questions; but what I want to sensitize you to is the importance of probing, really probing deeply, and really analyzing to insure your success during the later stages of the sales cycle.

8. Then, if you're selling systems, move to internal traffic. Find out which are the busiest departments, who communicates with whom, why and if there are any bottlenecks? How do they communicate after hours and across time zones? What are the priorities of the communication need, i.e., mobile, wireless, cellular telephone? Ask about growth, expansion, or downsizing?

9. For those of you selling long distance your analysis should be designed to fully understand their current service. Perform a complete analysis of their needs, then you can determine where you can help them, which is usually in the area of reducing telecommunications expenses. You have to determine their existing carrier, the monthly cost, the total minutes, or hours of traffic, the average cost per minute, and any additional service charges for all traffic (interstate, intrastate, international), calling card charges, 800 number calling, and directory assistance.

10. If appropriate, get a copy of the letterhead or company logo for use in your presentation later. Get a copy of the in-house directory for correct spelling of names and a telephone count.

11. The purpose of your fact-finding (assuming that you've previously done all the qualifying) is to be able to write out a list of your prospects' "area of concerns."

12. I suggested early on in this book that after your first appointment or conversation you should send your prospect a letter thanking them for their time, telling them about the next step, and when you'd be calling again or the agreed upon next step (meeting). Now, I suggest a second letter. It's

always important to stay in touch, it gives you a reason to call back, and it allows you to keep building credibility with written materials. In your second letter you should thank them again for their time, reiterate the next step (presentation, demo, or call back with solutions) — the time you start firing your gun — state a list of objectives for your next meeting, and attach a list of their areas of concern. In doing so you're demonstrating your full understanding of their telecommunications problems, and staying in full control of the sales cycle. You also will have a common ground upon which to talk...a mutual "frame of reference." The whole time that you are going though this fact-finding stage, remember our quote:

"Prescription without diagnosis is malpractice."

You can analyze, diagnose, probe, but give no solutions until you've fully completed the analysis, fact-finding stage...if you fire your gun before you see the white of their eyes (too early), you'll lose.

Look for Problems You can Solve

Here are a few ideas to remember during this fact-finding stage:

"You have more fun and enjoy more financial success when you stop trying to get what you want and start helping other people get what THEY want."

"When you want to remember how to sell you simply recall how you and other people like to buy."

"Remember...people don't buy your products and services...they buy how they imagine using them will make them FEEL."

Keeping the above in mind during the fact-finding stage, they will help you stay on track and remember what you're trying to accomplish. You're looking for problems you can solve with your products and services. But, how do you know when the prospect has a problem?

Okay, put down the book and either mentally or, if convenient, write down a list of the variables that need to be present or impacted before a prospect would buy your products or services.

This exercise is designed to focus you on "what's bothering your prospect" — not on what you can do for them. But don't forget that knowing something about your prospect is as important as knowing everything about your product.

Make a list of the variables or factors that need to be present or impacted for the prospect to buy your products or services. I can promise you with 100% predictability that one of the following things have to be present or impacted before your prospects buy anything:

- a desire to improve profitability
- a desire to increase revenue
- a desire or need to reduce expenses
- a need to increase profits
- a need to improve productivity
- a desire to reduce inefficiency
- a requirement for more safety
- a need to improve security
- a need for more stability
- a desire to improve their corporate image
- a need to provide better customer service

How many sound like some of the benefits your product and services can offer? Okay, now turn your thinking around and find out the areas of concerns based around the variables in the above list.

Once again, your objective during the fact-finding stage is to develop a list of areas of concerns. Your questions should be designed to lead the prospects to self-discover their areas of concern.

Questioning Guidelines

Here are some questioning guidelines to help you improve your fact-finding effectiveness:

1. Have a plan. You should know prior to the fact-finding stage exactly what information you must have in order to help the customer. It's always easier to have a roadmap outlining the information you're going to need during the call. If you don't have a good outline to prompt you as to the information you need to gather, design one.

2. Know what to do with the answers. Never ask a question when you aren't prepared for all the possible answers.

3. Be able to explain reasons for your questions. Always be ready to answer in terms of benefits. For example: "The reason I'm asking [the question] is that

I want to make sure you're getting the best rate for the high volume of calls you're making between New York and Boston, and determine if there is any way I can help you."

4. Explain your reasons for asking. Sometimes you will need to explain your reasons for the question, before you ask the question. Such as, "the reason I'm asking the next question is that it will help me determine how I can improve your customer service. Do you ever get complaints from your users that they can't get through during busy periods?"

5. Be conversational. React to your prospect's answers, respond, and build upon them. If you read your questions like a survey your customers will feel like a prosecuting attorney is interrogating them. Your questions should be mastered so that they come out naturally. This way you can focus on listening to your prospect rather than concentrating on what you are going to say next. When you make your prospect feel comfortable, they will open up and provide you with valuable information.

7. Use feeling questions. Which means that when someone is asked to explain how they feel about something, emotions get involved, and they are more likely to respond with an answer of substance.

8. Know when to quit. When in doubt, stop. It is better to be too brief than too long. Always end your meeting before the prospect gives you a look like it's time to end. But when you get really good at this, and I mean really good, you can purposely leave out a valuable piece of information, which will give you a very specific reason to call the prospect back.

RE-CAP

That's the gist of fact-finding. It's the best I can share with you regarding this critical phase of the sales process. When completed successfully, you'll be well armed for the next stage in the sales cycle — the presentation, demonstration, or the phase where you present your findings or make recommendations to your prospect.

I am concluding this chapter with a short discussion of success, plus a reminder to you to remember who is responsible for your success.

Is it your sales manager? Is it your general manager? Is it your spouse or loved one? Is it the president of your company? Is it the people that develop the capabilities of your product and services? Is it me? No, NO, NO! There is only one person responsible for your success — YOU. How do you know when you're successful? You don't. It isn't some place that you arrive — success is the journey, the process. Knowing that your reading this book encourages me to reassure you that you're well on your way — well into a successful journey.

MOTIVATIONAL SECTION

I want to end by sharing with you The Garden for Success. It's an idea that I've embellished a little, but it was inspired by D J. Harrington, a fellow member of the New England Speakers Association.

First plant 7 rows of Peas.

Presence

Promptness

Preparation

Patience

Perseverance

Positive action

Professionalism

Then plant 8 rows of Squash

Squash gossip

Squash indifference

Squash indecision

Squash negativity

Squash worry

Squash envy

Squash greed

Squash fear

Next plant 9 rows of Turnips

    Turn up for training

    Turn up on time

    Turn up with a smile

    Turn up with good thoughts

    Turn up with new goals

    Turn up with new prospects

    Turn up with excitement and enthusiasm

    Turn up with a positive attitude

    Turn up with determination

No garden is complete without Lettuce.

    Let us be honest with ourselves

    Let us be unselfish and loyal

    Let us be true to our obligations

    Let us have fun and enjoyment

    Let us obey the rules and regulations

    Let us love and help one another

That's the garden for success. So until we meet, happy planting, happy watering, may the sun shine, and especially happy selling. Go get 'em!

Step 5: Effective Presentations

Congratulations, you're over half way through. You've completed the fact-finding stage where you learned about gathering information about your prospect. And where I cautioned you to not give away your solutions until you reach this stage — the presentation or demonstration stage. You are chock full of information to share with your prospect. You have found solutions to their problems, and now is the time to share that information.

One of the goals of the previous stage in the sales cycle (fact-finding) was for you to be able to make a list of the client's "areas of concern."

One of the first pieces of business you will learn about your presentation will be for you to gain agreement with your prospect that the presentation covers those areas that are indeed within their areas of concern. You will learn how to solve the prospect's problems during the presentation process.

First of all, however, let's discuss a "truism" about telecom presentations. I've heard this from the top salespeople in the most successful companies in the Telecommunications Industry, and it doesn't matter whether they are equipment companies or services companies.

While it is not always possible or practical to have a prospect come into your office for a presentation or demonstration, when they do you can close more than 90% of the time. In other words, to increase your chances of making a sale to over 90%, all you have to do is convince your prospect to visit your offices. The way I see it, that is certainly easier than convincing them to part with their hard earned dollars.

One competitive local exchange carrier (CLEC) that I worked with recently confirmed that their closing ratio went up over 20%, when they began to show clients the Lucent #5 ESS switches located at or near their branch offices. Their prospective customers' most frequent objection was their fear of the products' reliability and staying power. The salespeople were able to address the customers' fears by showing them the company's five million-dollar investment in the community.

One senior sales executive put it this way. When they come to your office, they come to <u>buy</u>. When you visit them in their office, they are waiting to be sold.

There are, however, a number of things that you must do before you can assure yourself of a 90% closing ratio (or a 20% improvement in your closing ratio).

IMPORTANCE OF AN OFFICE VISIT

First of all, you have to convince your prospect of the importance of making an office visit for a tour, presentation, and demo. Here's how you do that.

From the beginning, sell as if it is a partnership arrangement. You are not a slave to your customer, wherein every time they ask you to do something for them you simply say "ok."

Whenever they ask you to do something during the sales process, you should, of course, agree — especially if it will make their decision easier — but in turn, ask them to do something for you. One of those things might be to get their agreement to visit your office so they can meet with the service personnel or see a live demonstration. Obviously, here I am directing my comments mostly to those of you that market telecommunications equipment or sell local services. But the equivalent of this for those of you selling long distance services or other telecommunications services is a personal visit by you to your prospect's office.

Now is a good time to put the book down and take a moment to consider whether a prospect's visit your office, where you can close the sale in the office is practical and/or appropriate for your particular telecommunications products and services. If the answer is yes, then start getting excited about a wonderful, extremely natural closing process. If the answer is no, you can still adopt the process outlined in this chapter, but just take it to the point where it ensures your success when giving the presentation in their office. Or perhaps where you can close over the phone when you present your solutions to your prospect.

THE PRESENTATION OR DEMONSTRATION

In this chapter I'll discuss presentation/demonstration — the stage of actually presenting your findings and giving your solutions to your prospect. Many of the concepts you will learn, apply if you're giving a full blown presentation and demonstration of your equipment to your prospect, or visiting them in person in their office and presenting the findings of your telecommunications analysis, or even presenting your findings to your prospect over the phone. As I share specific things you can do to ensure your success during this stage, I need for you to pick and choose those concepts that apply to your specific situation.

Think about the last time that you gave a presentation or demonstration of your products and services. If you think about it for a minute, you will probably concede, that at no other time during the sales process was the excitement for your product and the enthusiasm higher. However, as many salespeople have related to me, after the client's visit to their offices for the presentation, the enthusiasm took a slow downward spiral throughout the rest of the sales process.

Selling, in many ways, is a "transfer of enthusiasm." The slow deterioration in enthusiasm that is experienced after the presentation stage is highly detrimental to your success. You must ensure that your close comes when the enthusiasm is highest. This chapter is designed to share with you a process that will practically assure your success when it comes to getting an order at the conclusion of your presentation, and if appropriate, demonstration.

In order to prepare for an effective presentation, there are six essential steps that need to be completed during the fact-finding stage:

1. Get a copy of their letterhead.

2. Get a copy of their telephone directory.

3. Gather a package of their company information.

4. List their telecom areas of concern.

5. List their telecom objectives.

6. Layout a time frame and corporate visit date.

Step 1. A copy of their letterhead or logo is useful as you prepare your presentation since you can scan it into a PowerPoint presentation, or use it on a flip chart to get their attention.

Step 2. A copy of their in-house telephone directory with names and extension numbers helps you establish the memory patterns necessary to remember essential names and titles. Also use it to customize your live demos by using your prospect's real names and titles.

Step 3. Gathering a public relations file on your prospect's company consisting of articles on the prospect such as recent developments, newsworthy events, product information, or financial results helps you to gain respect, since the more you know about them, the more they respect you.

Step 4. During the fact-finding stage you asked, probed, and explored your prospect's telecommunications problem areas. You diligently took notes and from those listed your prospect's areas of concern. You then either mailed, faxed, or e-mailed the list to the prospect, or perhaps you discussed them and reached an agreement with your prospect on what their specific areas of concern were.

Try this now. Think of a prospect you're working on and write out as clearly as possible three areas where they have telecommunication concerns. This could be as simple as "concerned about rapidly rising telecommunication expenses" or "concerned about heavy traffic volume between Boston and NY" or "concerned about the bottlenecks at the switchboard, resulting in lost calls and potential orders."

Step 5. Make a list of your prospect's telecom objectives, such as: "reduce telecom expenses by 15% by years end" or "analyze long distance patterns, especially high traffic between Denver and San Francisco, evaluate other solutions, and take advantage of bulk rate potential savings" or "ease the bottlenecks at the switchboard to improve customer service and order flow." You've discussed with your prospect their telecom objectives, but you've been careful not to give any solutions, you simply demonstrated your understanding of their problems and goals.

Step 6. At some point during the fact-finding, or possibly qualifying stage, you met with your contact and jointly developed a time frame for their decision. Working backwards from the date of installation, cutover, or implementation date, you set target dates for the prospect to call your references, visit other competitors, get proposals, and you began planting the idea of a visit to your office. If you meet with resistance at this stage — good — because it was a valuable part of qualifying.

In a moment we will discuss how completing these six steps during your fact-finding stage apply during the presentation stage. However, first let's just assume that your prospect is coming to your office for a presentation, and/or demonstration.

PRE-OFFICE VISIT CHECKLIST

Here is a checklist of things you must prepare, prior to the visit, in order to ensure your success.

Invitations

Make sure you've invited the decision-maker and any other supporting cast, such as those that will assist in the final decision.

Have the Prospect see the Competition

You have asked the tough qualifying questions to determine your competition, but you've also set the stage for a presentation. In fact, in some instances, to demonstrate your confidence in your company's products and services, you've encouraged your prospect to visit other competitors. You have, however, always gained assurances that they will not make a decision until they've seen your live demonstration, or heard the results of your findings. I believe strongly that while it is important to be in the position to come into a company early in the decision making process to precondition the thinking, or simply educate your prospect, I also believe strongly that you will want to present last.

Always assume there is competition, and demonstrate your confidence in your products and services by asking your prospect to commit to visiting your office only after they've seen the competition. Don't misunderstand me. I don't want you to lose control of the sale — you must stay in touch with your prospect so you know who they are visiting, when they are visiting them, and encourage them to give you feedback on what they liked and disliked about what they saw or didn't see.

Have the Prospect Call Your References

The best way to convince anyone to go ahead with your recommendation is through testimonials. Let your best customers do the selling for you. It is important, in order to build credibility, to ensure that your prospect has called your references.

Here are some of my thoughts on references:

- This should be an open discussion between you and your prospect.

- Is it important that your prospect speak to other customers in the same industry?

- Is it important that your prospect speak to other customers with the same application?

- Is it important that your prospect speak to an individual at the same level in the organization?

- If your references are a specific 2 or 3 accounts that you know personally, you should call them and gently ask for permission to have someone call them, and then prep them for the call.

Maybe some of this seems pretty basic, but I've seen huge sales that have been won or lost specifically on the quality, timing, and/or outcome of a reference check. This is part of the closing process, so pay a lot of attention to the details, right down to your prospect asking your satisfied customer appropriate questions, and prepping your customer to respond appropriately. Be diligent in your follow-ups with your prospect to ensure that they got through to your reference, and the answers were satisfactory.

As a salesperson, one of your long term goals should be to develop your own personal list of satisfied customers, letters and referrals that you can use immediately, or take with you wherever you go.

Do a Trial Close

Assuming now that the prospect has enthusiastically done their homework — checked references, seen the competition, agreed on your time frame for making a decision, and also is aware that you've done your homework - you're in a perfect position to try a trial close. By the way, how do you think the prospect knows that you have done YOUR homework? In other words, how does the prospect know that you really understand their business? It's easy, it goes back to something you did during the fact-finding stage, you fed back to your prospect their telecom objectives, and you fed back to the prospect their areas of concern.

The trial close should go something like this: "Great John...so we can confirm that you'll visit our offices, take a tour, and see a demonstration, next Tuesday, July 13th. Do you mind if I ask you a quick question? In terms of everything we've been working on together so far, if during that visit, we can make you feel totally comfortable with our service and support, and if during the demonstration, you are convinced that the product meets your objectives, and if our price is competitive, is there any reason we won't be able to go ahead next Tuesday?"

If the answer is yes, and the prospect gives you a reason, then you've got some more preparation work to do prior to the presentation. If, however, the answer is no, then you're in a terrific position to close the order next week.

By the way for those of you that don't give demonstrations, but maybe just invite customers into your place of work, or go out and visit accounts, a variation of the trial close will work. Just ask yourself, anytime you are getting ready to visit with an account during the sales cycle, if you've listed that account on your business forecast sheet. If you've forecasted the business to close and you have not tried a trial close, or if you're afraid to do it, then you're just kidding yourself. If you do it, and the answer is "yes, there are reason why we can't close tomorrow (or next week)," then at least you can get those objections out on the table.

Let's say that after your trial close, the prospect says that one of the key decision-makers won't be available. I'd suggest postponing the date. Why? I'll say it once again — at no time will their enthusiasm to buy be higher that at the conclusion of the office visit.

Seven questions to ask yourself prior to inviting a prospect in for an office visit:

1. Is the decision-maker coming?

2. Have they seen the competition?

3. Have they called your references?

4. Have you done a trial close?

5. Have you scheduled the demo room?

6. Have you prepared an agenda?

7. Have you prepared your visuals?

Now is the time for you to make sure your entire company is prepared for the visit, and do the preparation for a company tour.

Book Demo and Presentation Room(s)

It is up to you to make sure the room(s) are clean and neat and then the demo must be set up. Here's where you use your prospect's in-house telephone directory because, if appropriate, you will customize the demo with their names and extension numbers. This is simple, yet critically important stuff. People like to feel important. By using their names in, for example, the speed dial demo, or the name-dialing example, they feel important and you demonstrate your commitment to understanding their business.

Let Your Company Sell Itself

Remember now, during this stage of the sales cycle, although you are selling yourself, in terms of your ability to understand their business, you are also giving a professional presentation/demonstration. But now is also an opportunity to take a little pressure off of yourself, and let your company sell itself. To do that you need to involve other people in the sales process.

How do you do this effectively? Visualize yourself like the conductor of the Boston Symphony orchestra.

You, the salesperson, are responsible for orchestrating a well-tuned home office visit and demo.

As the conductor you're responsible for the overall coordination of the orchestra, and you determine when and what each segment of the orchestra plays, how loud, what key, and so forth.

As the sales rep, you are responsible for the overall coordination of the sale, you, and only you, know if the prospect is concerned about service, or support, or the details of the potential installation, or the specific product or service application.

When you bring the customer into your office, you have the opportunity to introduce them to your manager, the service manager, possibly the owner of the company, the dispatcher and so forth. Each of these people can help you sell the company, but you need to prep them as to the prospect's needs.

Many successful reps write out a short summary of the sales opportunity, including the name of the prospect, the location of the account, the specific product being proposed, unusual applications, and hand delivers this information to those that will be involved in the on-premise visit and selling situation. Then the rep actually visits those that will be involved.

The day before you take a prospect on a tour of your company, you should visit the service manager and prep him/her on your prospect, acknowledge that you'll be introducing them the next day, and give a brief summary of the prospect. Then when you come by to make the introductions, the service manager can get up, shake hands, look them in the eye, thank them for stopping by, and ask them if they have any service oriented questions, and respond accordingly.

I can't stress enough what a favorable impression it can have on your prospect when they realize that the service manager has been told about the visit, and briefed on their company and situation. Service managers have a high credibility rating with most customers. Therefore, in some cases, it might be appropriate for the service manager to handle a small portion of the office visit, i.e., the warehouse/inventory, dispatcher, service order dispatch tour. It might be advisable for the service manager to show the prospect how their service order will be handled. As the sales rep, you must decide what company resources will be of help to you with a particular account: executive, management, operations, service, administration, or sales management. You're like the orchestra leader — choreographing the event, bringing the appropriate resources to bear on the situation, always coordinating and filtering the information that gets to your prospect. You must stay in control at all times, and the only way to do this is to brief those that are going to interface with your prospect and, if necessary, tell them exactly what to say. I stress this here, because I recently spoke with a sales rep that lost a sale because of what a sales engineer said to his prospect.

Prepare an Agenda for the Office Visit

This sheet will clearly have a list of attendees on it — the personnel in your company that the client will be meeting and the prospect's entourage. The written agenda serves two purposes: Your resources get to see the names of the visitors and can memorize them beforehand. Then you give it to your prospect so they can take with them a written record of who they met.

For a period of two years when I was marketing to the Regional Bell Operating Companies, which involved complex sales with long sales cycles, I had a sales manager that insisted that I make out an agenda for every client visit. After each visit, I would put the agenda in a file. There were times when a client returned a year later or so and I could pull out that agenda and tell them when we met previously, who we met with, and what was discussed.

The agenda also becomes one of your selling tools. It should have your company's name in a prominent position, the date of the visit, the time breakdowns down the left side and the corresponding item next to it. If different people will be responsible for different parts of the agenda, I suggest you list their names in a third column. It will be the first thing you review during your presentation.

Some suggestions for an agenda:

- The first item "Welcome and Introductions," allow 15 min.

- The second item " Company Tour including Service Dept.," allow 45 min.

- Item 3 should be a Review, first of the prospect's objectives (prepared and agreed to prior to the visit), and then their areas of concern, allow 15 min.

- This should be followed by a short Question and Answer period, allow another 15 min.

- Next should be System Demonstration allow 30 min. to 1 hour,

- This should be followed by your Presentation, Review of Proposal and Price Discussion, allow 30 to 45-min.

- The last item will be "Contract Details and Paperwork," allow 30 min.

Your time frame obviously will vary based on the prospect's needs and the complexity of the sale. I recommend, however, that the total time fit into a maximum of the 3-4-hour morning time frame, anything longer will lose its effectiveness.

SALES TIP:
Sample Office Visit Agenda:

Welcome and Introductions	15 min
Company Tour	45 min
Customer's Objectives	15 min
Question and Answer	15 min
System Demonstration	30 min
System Overview	45 min
Contract Details and Paperwork	30 min

The importance of the preparation of an agenda and its inclusion in your presentation can not be minimized. It allows you to put the prospect at ease at the very beginning of the presentation since they can see exactly what will happen. It helps to keep everyone on track, and most importantly, it allows you to open up the close. The prospect will clearly see right at the beginning of the day, that at noon, or at the end of the day, you will be asking for the order. This should come as no surprise, if you've done an adequate trial close prior to the visit.

Prepare Your Visuals

I speak to large audiences, give presentations, and train salespeople for a living. Of all the visuals aids I could use, such as PowerPoint presentations from my laptop, slides, overheads, flip charts, I prefer to use either an overhead projector combined with flip charts, or PowerPoint presentations. Slides are flashy, expensive, and usually require that you turn the lights down and many times this limits an interactive discussion. I prefer to reserve slides for my larger, more formal presentations. Laptops and PowerPoint presentations are more flexible and easier to customize for your audience. The computer projectors have been perfected so your visuals can be seen with the lights on. Overheads are easy to make (especially with the great new graphics-oriented software) easy to print out and copy onto overhead foils. Another reason I like overheads and/or PowerPoint presentations is because they allow you to stay on track and keep the presentation focused.

You might ask, why use visual instructional materials at all in your presentation? General presentation studies show that if you merely "tell" someone something, 3 hours later they have retained only approximately 70% of the information, and 3 days later that audience will have retained only 10% of what you told them.

However, if you "show" the audience something, 3 hours later they will have retained 75% of what you showed them, and 3 days later they will have retained 20% of the information. But, if you both "show and tell" your audience something, they will retain up to 85% on the information 3 hours later and 65% of it 3 days later. Therefore, make it a habit of using visuals with your presentation.

SALES TIP:
Why you should use Visuals in your Presentations:

Medium used	Retention-3 hrs	Retention-3 days
Tell only	70 %	10%
Show only	75%	20%
Show and tell	85%	65%

What should be on your visuals as you prepare for this important office visit? I suggest a very simple format.

Visual 1 - The Welcome. It should be a simple welcome, such as, XYZ company welcomes XYZ company. Here is where you can get creative, either on an overhead, on a flip chart, or in PowerPoint, copy, draw, or scan in the prospect's logo. Recently, I've been simply looking up my clients' Web sites and cutting and pasting their logo into my PowerPoint slide. During all the years I've been selling, this approach always seems to win me some points. With the overhead or visual up you can begin your presentation, which is usually the most awkward part of the presentation, but it also must to be the smoothest, spiced with a little levity and a sense of humor — something that attracts attention.

The best openings I've heard are usually heartfelt comments about commonalties between the two companies, something about the event itself, or about the situation at hand. Whatever your opening comments, don't try to wing it. You must develop on paper your remarks and practice them, so that they come out smoothly. Move quickly into thanking them for their time, doing the introductions, and then tell them what you are going to tell them by putting up the agenda.

Visual 2 - The agenda. Cover the agenda, getting agreement on each segment and agreement with the time frame for the morning or afternoon, and most importantly you will open up the close again, by pointing out the time segment for the completion of the paperwork. By the way, I have seen situations where the customer says something like, "You know, this day is really becoming quite busy, can we go right to the paperwork?" In such a situation, I suggest that you skip the rest of the agenda, and get out your contract.

Visual 3 - A list of the prospect's telecommunications objectives. Review and get agreement that these are their objectives and state clearly what your process will be to help them achieve their objectives. I like the idea that, if there are multiple objectives, you leave a space or box to actually check off each objective as you get an agreement.

Visual 4 - A list of the prospect's areas of concern. It is probable at this point that you will leave the presentation and move to the demo, and in a moment I'll cover some suggestions for effective demonstrations. In any event this visual should allow you to prep the prospect for what they will see in the demo. At the completion of the demo, you can come back and refer back to these areas of concern.

Visuals 5 and 6 - Features and benefits. This should be a list of probably no more than three-five major features of your system set out on the left side, and the benefit of that capability inserted on the right side of the overhead or flip chart. Remember, typical telecommunication benefits are not too complicated. In brief, benefits either:

- save time
- save money
- increase revenue
- decrease expenses
- improve productivity
- improve effectiveness
- improve corporate image
- improve customer service
- are safety or security oriented
- increase market share
- improve time to market

A good exercise is to think of four or five features of your equipment and relate the appropriate benefit to that capability.

When giving your presentation, make sure to always involve your prospect by asking such things as: "So can you see how that would improve productivity?"

"Do you see how much that could help your corporate image?" "Can you see how that could increase revenue?"

If your prospect is operations oriented or accounting and financially-oriented (and it's a good guess that they are if they're the decision-maker, or perhaps you've been told, "he or she is a by the numbers manager"), then you should have financial back up to your presentation. This could include any payback analysis, a cash-flow analysis, or a return on investment analysis. If your prospect is looking for financial justification, and you don't give it to them at this point, you will fail. And the only way to know if it is needed is to have asked.

As I stated in Chapter 1, one of the aspects I love the most about selling telecom equipment and services is that the products and services are usually 100% cost-justified. Use this principle where and whenever possible. If there is a lot of detail involved, I suggest you handle it on a separate piece of paper as a hand out. Make sure the hand out clearly outlines the assumptions behind the figures. You can mention that you have a financial analysis that goes into the figures, but I suggest you hold back the discussion of finances until the end.

One more thought with regard to doing a financial analysis with your prospect. Isn't it always better when the prospect does this with you? Isn't it wonderful when the customer shares with you their figures? The best situation is when the prospect says things like "Here's what we need for a return on investment?" or "By my calculation, our pay back looks like around three years." Having a prospect mutually develop the cost justification with you is a sure sign that you're on track.

<u>Visual 7 Cost justification and paperwork</u>. This is the last visual and should clearly set the stage for the close. It could simply say contract negotiations, or completion of paperwork, or review of proposal, financial analysis, and contract. Basically, it allows you to now pass out the proposal, hit the highlights, and end on the contract page. I don't want to make this sound too simple. At this point you should have full agreement that your prospect is comfortable with your service and support, the product answers any concerns and it meets the prospect's objective. So, as long as your price is competitive, this should be the close.

RE-CAP

Let's briefly review the seven steps you must take prior to inviting the prospect into the office.

1. Make sure the decision-maker and any supporting cast is coming. The worst thing that could happen to you would be at the end of the presentation, you hear something like this: "Well let me pass this by Mr. Big." Don't let that happen to you.

2. Insure that they've seen the competition, (all the competition), and that you are the last. Prior to their visit to your company, you've continually checked with them to see what they liked and disliked about the competition.

3. Insure that they've called your references. These were not just any reference, but the kind of references that they had requested.

4. You've completed a trial close. The timing of this was crucial, as you did it when they were upbeat, excited about you, your company, and optimistic about working together. When you did it, it seemed like a natural outgrowth of the existing conversation.

5. You've ensured that the demo and conference room is available and booked, and everyone that will be involved in the visit has been briefed as to who's coming, why, and what their role in the sale will be.

6. You've prepared an agenda and passed it out to those involved. You felt comfortable with the last item on the agenda — completion of paperwork, because you've set the stage for a close that day.

7. You've prepared your seven visuals — either as overheads, PowerPoint slides, or flip charts:

    Visual 1   Welcome

    Visual 2   Agenda

    Visual 3   Their telecom objectives

    Visual 4   Their areas of concern

    Visuals 5 & 6 Features and benefits

    Visual 7   Cost-justification & paperwork

YOUR PRESENTATION SKILLS

The last item of business is to prepare **you** for the presentation. This can be either very simple or possibly the toughest thing you've ever done in your life, depending on your level of experience.

We do know that of the five things that people fear most are:

- going to the dentist
- an operation
- flying
- dying
- public speaking

Let me put it as simply as possible — to be effective in this great wide wonderful world of telecommunications, and to really make it in the real world, you should begin today to perfect your skills as a speaker and presenter. Our industry has allowed thousands of salespeople to perfect their speaking and presentations skills in front of relatively small audiences before venturing out in front of larger groups. So if you're not fully comfortable with standing up in front of three, four, or five people at the end of a conference table, this is a perfect chance to learn and perfect you speaking skills. There are many wonderful books published on the subject, and I know many people that have benefited from involvement in their local toastmasters group, or any number of in-house Presentation Skills training.

I should mention here that I conduct a one or two-day Presentation Skills class, for up to 15 attendees. I video tape the presenters in a thoroughly non-threatening environment, and everyone has a lot of fun perfecting their speaking skills.

We work on all the different aspects of speaking, including how to hold your hands, how to avoid using "ums" and "aahs", how to stand, use eye contact, how to get organized, how to prepare your visuals, how to get your audience involved, and much, much more.

The course is based upon the material in Claudyne Wilder's two books: *The Presentations Kit...Ten Steps For Selling Your Ideas*, and *Point, Click, and Wow*.

Here are Claudyne Wilder's ten steps:

Step 1 How to channel your nervousness.

Step 2 How to define your objective.

Step 3 How to fully organize your presentation.

Step 4 How to create and use effective visuals.

Step 5 How to energize yourself.

Step 6 How you can motivate your listeners.

Step 7 How to conclude with conviction.

Step 8 How to manage questions and objectives.

Step 9 How to strongly recommend the next step.

Step 10 How to take the leap into excellence.

I have known and worked with Claudyne for years, she is a true professional and has built a wonderful speaking, seminar, and training business. She runs the entire business from her headquarters located in Boston, Massachusetts. You may contact her at Wilder Management Services, 57A Robinwood Ave., Boston, MA. 02130, Phone and Fax (617) 524-7172. Claudyne's clients include such prestigious companies as Gillette, Fidelity Investments, and IBM/Lotus. For a number of years, I've taken her material and customized it for the Telecommunications Industry.

Claudyne would insist that I take a few pages of this book to make some suggestions you should keep in mind when presenting.

What Your Audience Wants
What do you think is more important during a presentation...the speaker or the content? Research show that people remember the speaker more than the content. The audience members are saying to themselves: "Can I trust this person?" "Can I believe this person?" In a selling situation, your believability becomes critical. "Can I trust this individual to understand my telecom needs, interpret them clearly and articulate them to his or her organization in order to fulfill my telecommunication requirements?"

When giving your presentation, try to remember what the audience is looking for. From experience, we know that your audience is basically looking for six different things that either add to your credibility, or erode your credibility

1. <u>They are looking for a strong opening with wit</u>. As I mentioned earlier, write out and practice your opening. The first minute of your presentation is the most important; you can't afford to stumble around.

2. <u>They want you to look at them</u>. Look them right in the eyes. It is impossible to look someone in the eyes and lie. There is a joke that the only people that can do this are teenagers, psychopaths, and lawyers. That is why it is critical to look your prospects right in the eyes when presenting. This takes practice, but is an important part of your learning process. The prospect must feel totally comfortable with your ability to help them.

3. <u>They want your voice to have a sincere tone</u>. Your voice, including how you say things, your tone, your modulation, how sincere you sound, your pace, all add to your credibility and believability. Tape yourself and listen to yourself. Try to get the pace, tone, modularity and sincerity correct.

4. <u>They want movement</u>. Movement around the room is important. Don't stand in one place, and never ever use a podium to stand behind or hide behind.

5. <u>They want to hear your message loud and clear</u>. Try to develop one theme or message for your entire presentation. And whatever that message, repeat it over and over again during the presentation. If it's "easy to use" say it many times. If it's "quality" repeat it often, if it's "local service" say it.

6. <u>They want to hear, see, and get involved in what you are talking about</u>.

Here are my ten suggestions for giving an effective presentation.

1. Develop a strong opening.

2. Look them in the eyes.

3. Vary your voice.

4. Show them your body.

5. Move around.

6. Have a single theme or one message.

7. Use visuals.

8. Ask questions - get them involved.

9. Tell short stories, use examples.

10. Practice.

Well, there you have it, an absolutely proven effective process for closing 90%

of the prospects that visit your office for a presentation. If you follow the process, I guarantee results. And for those of you that believe your sales force needs to further perfect their presentation and speaking skills, give me a call and we'll talk about the Presentation Skills course.

MOTIVATIONAL SECTION

Success

There are no secrets to success.

Success is doing the things you know you should do.

Success is not doing the things you know you should not do.

Success encompasses all the facets of your relationships as parent, spouse, citizen, neighbor, and worker.

Success is not confined to any one part of your personality, but is making the most of your body, mind, heart, and spirit.

Success is discovering your best talents, skills, and abilities and applying them where they will make the most effective contribution to your fellow man.

Success is 99 percent attitude.

Success is a continuing upward spiral of progress. It is perpetual growth.

Success is relative, individual, or personal. It is your answer to the problem of making the minutes, hours, days, weeks, months, and years add up to a good life.

- Wilfred A. Peterson

Step 6: Overcoming Objections

"We spent our entire telecommunications budget...honest."

"That was a wonderful presentation...but I'd like some time to think this over."

"Good demo...but I'd like to talk this over with my people."

"Your price is too high."

"I'd like to wait till business gets better."

"Sounds great...but we're satisfied with our existing long distance carrier."

"Not right now...but can you send us some information on your company."

"We're looking at a number of other telecommunication solutions."

"We'd like to shop around."

"I'm concerned about your company's long-term commitment to the industry."

"Get back to me in about 3 months...we'll be ready by then."

After years of selling telecommunication products and services, just writing down the above objections makes me sick. The objections bring back painful memories of the times, especially when I was a rookie, when I seriously thought that I had a legitimate shot at closing a sale.

If you've ever been in a position where you believed that you had invested a lot of time and energy, and had earned the business you were seeking, only to hear one or more of the above objections, then Step 6 is for you.

In this chapter we will:

- Discuss the different types of objections.

- Learn how to recognize phony objections.

- Learn how to find out the real objection.

- Review the process of handling objections.

- Identify and practice handling the two most common objections in the telecom industry.

TYPES OF OBJECTIONS

The first step in understanding how to overcome objections is to first understand the five different types of objections.

Conditional Objection

This is probably the worst kind of objection you can hear. A "condition" exists that prohibits the prospect from going ahead with you. The best example I can share with you is when your prospect's company is about to be bought, or merged with another company, or maybe has fallen on hard financial times. You ask a closing question and the prospect says: "I'd love to go ahead with you today, but my hands are tied because of the upcoming merger."

This is a condition that neither you, at this point, or apparently your prospect, has any control over. You'd be well advised to back off, keep the door open for future business, but seriously think about finding a new prospect.

The Cover Up

Many buyers actually hide the true objection. All of my selling life, I've heard veteran salespeople say; "Buyers are liars." That's always been a little too strong for my way of thinking. But why do buyers hide the true objection? I think it's usually because they don't want to hurt your feelings, or they're embarrassed, or they're simply afraid to tell the truth.

Many times they might think that a "white lie" is easier than actually telling the truth. But many times they are just trying to get rid of you. When a prospect says something like, "business is slow right now" or "we've already spent our entire budget," you're job is to determine if this is true or if they really are just trying to get rid of you. If objections like these keep coming up at or near the end of the selling cycle, then you should start to rethink your qualifying capabilities.

The Stall

A stall is the type of objection that gives the prospect more time. When a prospect says something that "stalls" or "delays" the decision, they're usually saying something else. When you hear a prospect say "I'd like to think it over," or "I'd like to sleep on it," or "I never make decisions in front of salespeople," you are hearing good examples of stalls. But when you hear a prospect say any of these, they are probably buying time to do something else such as, look at a competitor's offering, find a better price, or check your references. Later we'll look at a process to find the "real" objection.

A Challenge

A challenge is usually an objection that occurs as a direct result of something that your prospect has heard from a competitor. One particularly large equipment distributor in our industry had emerged in a financially sound position from a Chapter 11 reorganization. They had reorganized and were now owned by a wealthy Midwestern conglomerate. They were in much better shape financially than they had been before the reorganization. When I started working with their salespeople, a full year after the reorganization, their financial health came up in almost every competitive sales situation. Invariably they would discover that the competition was bringing up the issue.

If your company has any weakness, such as a weakness similar to the one mentioned above, you can be assured that the competition will bring it up. When you, as a salesperson, hear what you believe is a challenge, be prepared for it, isolate the source of the challenge, and deal with it professionally.

A Question

A legitimate question is probably the best kind of objection you can get. It means the prospect has a valid concern about something you said, or is concerned about some aspect of your offering. A valid concern or question should simply be handled professionally and taken right to a close. Some examples might be does your price include the two-year extended warranty or, do your cost savings include an analysis of my third-party calling card traffic? These examples of legitimate questions and concerns are buying signals.

Of the five different types of objections — conditional, cover-up, stall, challenge, and question — the last two: a challenge and question — should actually be encouraged and welcomed. They are pretty easy to recognize. Try to view legitimate questions and challenges as steps toward a close. Maybe you've heard this before, but without legitimate questions, without challenges, you probably don't have a qualified prospect. On the other hand, the other three types of objections, conditional, cover-up, and stall are much harder to recognize.

When you ask a closing question in the Telecommunications Industry and the prospect voices an objection, it is usually not the "real" reason they won't buy. What they are saying is: "You haven't sold me yet." Usually the prospect needs more information or more assurance that they won't make a bad decision.

What you will soon learn is there are few "real" objections. Most are conditions, cover-ups or stalls. Your job as a salesperson is to ask the right questions to determine if there is a legitimate "condition" preventing a go-ahead, or if the prospect is lying to you so they won't hurt your feelings or stalling so they can find out additional information.

> **SALES MANAGER'S EXERCISE:**
> Assign each salesperson one of the five different types of objections. Have each salesperson present to the class the type of objection, how to recognize it, and at least three examples of that type of objection.

DAVID LETTERMAN'S TOP 10 OBJECTIONS

Recently, I heard David Letterman give the top ten reasons people give salespeople in order to avoid making a decision. Of course, David Letterman didn't call them "cover-ups" or "stalls", and he didn't tell you they were little white lies. He also didn't tell you they weren't the real objection, but you'll recognize them as such. Here they are in descending order starting with 10.

10. Our ad agency handles that.

9. Business is slow right now.

8. Quality is not important to me.

7. Get back to me in 90 days, we'll be ready by then.

6. I'm not ready to buy yet.

5. I never purchase on impulse, I always give it time to sink in.

4. I need to sleep on it.

3. I want to think about it.

2. We've spent our budget.

And the #1 reason (actually "stall" or "cover-up", or "white lie") people use to avoid making a decision:

1. I need to talk it over with my (partner, manager, wife, cat, mistress, accountant, lawyer, or my shrink).

Any of you that have been selling for awhile in the industry can probably make up a list of your own favorites. So, like you, I have my list. Here are the Philip Max Kay top 5 little white lies, stalls, or cover-ups I've heard, starting with #5.

PHIL KAY'S TOP 5 OBJECTIONS

5. We buy everything out of the home office.

4. We're satisfied with our existing telecommunications provider.

3. We always get three bids.

2. I need to talk it over with my management team.

And the #1 cover-up, stall, or conditional objection in the Telecommunications Industry:

1. You're price is too high.

THE "REAL" OBJECTION

Now you know that none of the above objections are the "real" objection. So if none of these are the real objection, what is the real objection? The real objection, the real reason that your prospect won't go ahead with your recommendation at that moment is usually never stated outright. When your prospect says: "I want to think it over." Or "I never make decisions in front of salespeople." Or gives you a stalling line like, "I need to sleep on it." He or she is really saying something else.

What are they really objecting to? What is the "real" reason for not going ahead?

I thought you would never ask. Here are the Philip Max Kay top 20 reasons for a prospect **not** going ahead with your telecommunications recommendations:

1. The company doesn't have the money, or hasn't budgeted for the money.

2. The company has the money, but doesn't see the value in spending the money.

3. The prospect knows they can't get the necessary credit approvals to finance the purchase.

4. The prospect can't make a decision on their own.

5. The prospect you're working with doesn't have the authority to spend the company's money without someone else's approval.

6. They think or know that they can get a better deal somewhere else or the same system somewhere else at a better price.

7. They have something else in mind, but won't tell you.

8. They have a friend or connection in the industry.

9. They like their existing telecom vendor, and do not want to change.

10. They want to shop around.

11. They are just talking to you in order to compare your price to the preferred vendor's price.

12. They are still confused about some aspect of your offering.

13. They are too busy with other more important things at this time.

14. They don't need, or they think they don't need your service or product at this time.

15. They either think or know that your price is too high, and/or always negotiate price with vendors.

16. They don't like or have confidence in your offering.

17. They don't believe you can provide the cost savings you demonstrate.

18. They don't have confidence or trust that your company can provide the service and support that you say you can provide.

19. They don't believe that the promised cost savings are worth the risk of lower quality, service or support.

20. They don't like, trust, nor have confidence in you.

As a salesperson in the telecommunications industry, part of your job description is to develop the capability to find the "real" objection. You will only be successful making the sale if first you find the real objection, then overcome it. You might be able to overcome the stated objection perfectly, but if it wasn't the "real" objection, you'll still be sitting there wondering why you didn't make the sale.

Here is a process to determine the real objection. You can overcome objections and move on to the close and in the process obtain the help necessary to get the sale. Notice that I did not say "guarantee that you get the sale." Why? Because in selling, like in life, there are no guarantees.

The following process will indeed dramatically increase your chances of turning a sale around. Fifty per cent of the time when you hear an objection like: "You price is too high," or "I want to think it over"; you're finished. Maybe you are finished. Maybe your price is really too high, and maybe they really do want to "think it over" and stall you long enough to have time to look at Lucent or Nortel or Sprint or Cisco.

But take heart, because fifty percent of the time when you hear an objection, you have a legitimate shot at the sale, and the prospect can be sold if you use the right process.

FORMULA FOR HANDLING OBJECTIONS

Hear them out: Listen carefully to the question or concern being raised. Look them in the eyes, don't interrupt, hear the entire objection.

Acknowledge their concern: No matter what, agree with your prospect at first. Say something like: I can understand your concern, or I had a customer last year with the same question before they decided to go ahead with our telecommunications system.

Feed it back: Try to feed back to them their question. Try to refer to their objection as a question. Sometimes by just feeding back to them their objection, you can get the prospect to amplify or better clarify their original objection. This is important, because the problem with an objection like, "I want to think it over" is that there is nothing to sink your teeth into. Simply by feeding back to them with something like: "So you want to think it over?" You might get a response such as: "Yes, I want to think about the two different pricing options I've received," or "the different service and support capabilities from the two different companies I am considering." At least now you have something you can address.

Restate the objection: Now you can restate their objection. Start by saying something like; "So, if I understand you correctly, your question is. . . ." or "It sounds like what your really trying to say is. . . ." or "If I understand your question correctly, it is — is my service and support superior to my competitor?"

Qualify the objection: Try to make each objection the last objection. You could say something like: "Is that the major stumbling block between an agreement and not going ahead today?" or "In other words, if it weren't for this you would probably go ahead today?"

Answer the objection in a positive way that incorporates the solution. You might say: "So if I was able to prove the superiority of our service offering over Winstar's competitive offering, would that be enough for you to make a decision?"

If your prospect really feels that this is indeed the only barrier to going ahead at this point, and if you satisfy their concern at this point, you will have a sale. The key to the sale is to have done your homework and preparation. And you need to have the proper tools in your sales kit (discussed later in this chapter) to win. For example, if a prospect is concerned about service — pull out a powerful reference from a similar customer that loves your service, or pull out testimonials.

If a prospect is worried about long-term commitment to the industry — pull out market share information, or a list of well-known national or local accounts that are doing business with your company.

If a prospect wants time to shop around for a better price — show them a cost-analysis or a competitive analysis, if you are offering any time-related deals or price-related deals. In other words, this is when you use every sales tool you have.

Ask another closing question: If, indeed, you have dealt with the final objection, and/or the last remaining true objection, then you should be able to conclude with saying something like — "So if that settles that issue, then can we go ahead with the paperwork?"

SEVEN STEPS TO HANDLING OBJECTIONS

I now give you a simple list (that you can keep in your PDA—personal digital assistant— if you like), which I call the seven formal sales steps to handling objections:

1. Hear them out. (Never interrupt, look them in the eye, and listen closely.)

2. Acknowledge their concern. (I can understand your concern.)

3. Feed it back. (Repeat, in their words, the objection.)

4. Restate the objection. (Restate or rephrase the objection — So if I understand correctly your question is. . . .)

5. Qualify the objection as the only true or last objection. (Is there anything else?)

6. Qualify the objection to set up the close. (If I can demonstrate . . .can you go ahead today?)

7. Ask another closing question. (So now can we go ahead with the agreement?)

OVERCOMING OBJECTIONS

Let's talk for a minute or two about objections in general, before I share with you some specific dialogue covering common objections in the Telecommunications Industry.

There are two things I want you to remember about objections:

1. **Don't fear objections**

Try to view objections as the natural progression towards the sale. Without objections, you probably don't have a prospect. For, without objections or questions, your prospect probably isn't very interested in your offering.

Objections open the door to the close. A valid and legitimate objection can open the door to a close, a close that is natural and without pressure. So welcome objections, handle them professionally, then move on. By now you know that to handle an objection professionally means:

- to hear them out
- acknowledge their concern
- feed it back to get clarification
- restate the objection to get further clarification

- qualify the objection as the last objection
- set up another close
- end with another closing question

2. Be prepared for objections

The major objections in our industry are predictable and heard often. In many cases it is preferable to even bring up the objections during the sales process. Bring up the typical objections before your prospect brings them up. Bring them up and practice answering them. The best way to be prepared for objections is to develop sales tools that enhance and support your ability to answer common objections.

Sales Tools

The following sales tools must be readily available so you can easily whip them out when you are faced with a prospect's objection.

Testimonial letters - these could be used to make a prospect feel more comfortable with your company's service and support.

Product comparison charts - these could be used to give a prospect rational reasons not to waste time shopping around.

Market share documentation - these might be used to give a prospect comfort regarding your company's long term commitment to the industry.

National account users list - These can help you prove value to overcome a price objection because major accounts are known for shopping hard for value.

Reference list - a list of companies or personal references that had the same concern as your prospect.

Financial analysis - possibly proving value by showing the real cost of ownership over 5 years.

SALES MANAGER'S EXERCISE:

Discuss with your sales reps sales tools that would be helpful in overcoming the most common objections. If applicable, assign a rep to develop a sales tool to share with the entire team.

Scenarios — Overcoming Typical Objections

Arm yourself with powerful sales tools, and have the right attitude toward handling objections. Let's look at a couple of different scenarios:

In the first example, I begin by asking a closing question, and the prospect, called Tom, will give me an objection about needing more time to think about it. Notice what I say in order to find the **real** objection.

Sales Rep: So Tom, I think that just about covers everything, can we go ahead with the paperwork?

Prospect: I don't think so...I need a little more time to think about it.

Sales Rep: I can fully understand that...there are a lot of considerations that go into a decision of this magnitude...so you need to think about it?

Prospect: Yes...I need to think about it. In particular, I need more time to compare your offering to a similar one that I received from Intertel.

Sales Rep: So, if I understand this correctly, the decision comes down to comparing and contrasting the capabilities of my system to the Intertel system.

Prospect: Yes, in fact the prices were just about identical.

Sales Rep: Terrific...so let me ask you this...if I could demonstrate clear superiority of my offering over the Intertel system with regard to capabilities that you need now, capabilities that clearly solve your existing problems, you'd be able to go ahead today.

Prospect: Yup...I guess so.

Sales rep: Well Tom, what I have prepared is an analysis of the top five capabilities you requested, clearly showing our superiority over Intertel...and here they are. Can you see that...and look at this...and in this particular area...we have clear superiority. Don't you agree?

Prospect: Looks pretty good to me.

Sales Rep: So, I guess that settles that...so if I can get your approval on the paperwork, we should be able to start your installations within two weeks?

Prospect: Well let's go!

Many of you would probably agree that Tom was pretty easy on me, and as you know most of the time it's not quite that easy. The important point to remember, however, is to master the process that I used. If you re-read the above scenario, you will see the application of the process we just covered. You can see how I overcame Tom's original objection, "I want to think about it," by getting Tom to amplify his original objection from "I want to think about it" to the specifics of what he was really concerned about. You can see how I was able to isolate his objection as the final or last objection and move toward another closing question.

Now let's look at how to handle one of the most common objections you will hear:

Sales Rep: So Tom, can we take care of the paperwork today?

Prospect: No, I don't think so...not today, can you call me next week and I'll give you my decision.

Sales rep: Call you next week for a decision?

Prospect: Yes, call me next week for a decision.

Sales rep: Tom, I don't mean to make light of this, but usually when a potential customer like you says something like that, they usually have a good reason for delaying the decision...do you mind sharing with me what it might be?

Prospect: Sure, it's your price.

Sales Rep: Tom, can I ask you specifically what it is about the price that concerns you?

Prospect: Certainly, you are $700 more than Comdial.

Sales rep: So, it's this $700 that you're concerned about?

Prospect: Yes!

Sales rep: So, if I understand what you're telling me, if it weren't for this $700 difference you'd be prepared to go ahead with me today.

Prospect: Yes, I guess so.

Sales rep: Tom, remember when we first met and we talked about price? I made it clear that we wouldn't be the **lowest** priced vendor that you could get a bid from?

Prospect: Yes, I recall that conversation.

Sales rep: Well I also told you that we'd probably also not be the **highest** priced vendor that you would receive a bid from...but related that we'd be within 10 to 15% of the major manufacturers. If you look at the total cost of our system we certainly are within 10% of that competitive bid...but, I know that when you're making a purchase such as this you want to make sure that you're getting the greatest value for your dollar.

I guess the real question in your mind might be something like: "How can you justify spending $700 more for our system over Comdial?"

Prospect: Yes

Sales Rep: Well a number of other companies I have worked with had a similar concern, but after becoming customers have discovered a lot of value in working with us.

Here is an outline of some of the special things our company does during installation.

We talked about service and support, but here are a number of letters from customers raving about our service department.

Here are three letters from customers raving about the same capabilities that you need in your system.

I also wanted to show you a list of local well-known companies, companies that are very demanding, that have seen the value in our offerings.

Lastly, when you really think about the $700 spread over the life of say 5 years, that's $140 a year, less than $12 a month, and less than 50 cents a day.

So if your question is: "How do I get the greatest value for my dollar?" Well, that's the same reason that you should invest in our system today, for the greatest value, all I need is your approval.

Prospect: You got it!

RE-CAP

Now that you have learned how to overcome some of the more common objections in our industry, and have reviewed a couple of examples of legitimate objections, I'd like to remind you once again about how to use this part of the sales process. I've known for years that the key to learning is repetition. So with that in mind, review the process, then practice out loud, over and over again, overcoming typical objections. If you have someone to practice with, that's all the better.

I can't wait for the first time you hear one of these objections in a real selling situation. You can take a lot of pride in the fact that you read this book, and you were introduced to the process, learned the process through repetition, and made the sale.

Do not fear objections, welcome them. Objections open doors to the close, and when you get the really tough objections, the key to your success will be your preparation — which starts now, as you think of the typical objections you encounter, and what kinds of documentation you need to have with you in order to overcome those questions and concerns. Your preparation will ensure your success.

SALES MANAGER'S EXERCISE:
Review the seven-step process for overcoming objections. List on the board the top three or four objections your salespeople hear. First, practice the two examples in the book, then practice using the objections on the board.

MOTIVATIONAL SECTION

Over the years, I've heard people say that it doesn't matter what happens to you. What does matter is how you handle what happens to you.

Because attitude is so important to success in selling, I'd like to share the following on attitude. I grabbed it off the wall of one of the hardest working, squared away, sales managers in all of New York City. It came from Art Herold, the sales manager of Gillette Global Networks, located at 39 Broadway in the big apple.

ATTITUDE

The longer we live, the more we must realize the impact of attitude on life.

Attitude is more important than facts.

It is more important than the past, than education, than money, than circum-stances, than failures, than successes, than what other people think or say or do.

It is more important than appearance, giftedness, or skill.

It will make or break a company...an organization...a home.

The remarkable thing is we have a choice every day regarding the attitude we will embrace for that day.

We cannot change our past...we cannot change the fact that people will act in a certain way.

We cannot change the inevitable.

The only thing we can do is play on the one string we have and that is attitude.

Life is 10% what happens to us and 90% how we react to it.

And so it is...we are in charge of our . . . ATTITUDE

And last is this little reminder:

Yesterday is but a memory
tomorrow but a vision
but a day lived well
is a memory of happiness and a vision of hope.

Step 7: Closing

"The first person that speaks loses...The first person that speaks loses...The first person that speaks loses!"

If you're a veteran salesperson, you've probably heard this before. Maybe you heard it from one of the old timers in our industry, or from another salesperson, or your first sales manager, or one of the traditional, popular generic sales trainers. Wherever you heard it, it really doesn't matter. What matters is that you are open to discussing closing philosophies, understanding some of the principles of closing, and specifically improving your ability to close a higher percentage of the qualified prospects you are working on.

So where does "The first person that speaks loses" come from?

The first person that speaks loses!

Apparently in some industries, in some sales environments, after a salesperson asks a closing question, and then shuts up, the tension builds to such a point, that the prospect is compelled to answer in the affirmative. It is thought that if the salesperson has answered and overcome every imaginable objection, and can wait long enough after asking a closing question, the prospect, out of some kind of exasperation (if the salesperson can shut up long enough), will be compelled to relieve the tension, by saying, "OK...let's go" or "OK...I'll do it," or "OK...I'll buy it" or "OK...$16,000 is a lot to pay for one-week a year in Cozumel, but the view is lovely, and it forces us to take a vacation every year, so OK" or "OK, I know $38,000 is a lot to pay for a car, but it's what we've always wanted, and I love the metallic blue color, and you've been so nice, and I'm tired of all this aggravation over buying a car, so where do I sign?"

This chapter is totally devoted to the art of closing, and obviously I'm being a little facetious, and having a little fun taking on some of the well known closing techniques of the Timeshare Industry and the Auto Industry.

Because I've been fortunate to have trained thousands of salespeople in the Telecommunications Industry, and have sold these products and services for many years myself, I am proud to say that I've personally witnessed a high level of professionalism in the Telecommunications Industry. Although the "first person that speaks loses" technique has its place, the contrived sales tactics, tricks, gimmicks, fancy words, unsubstantiated promises, charades, and many of the more elaborate closing techniques you may have heard of, or are aware of, just won't work in the Telecommunications Industry.

When a sales manager comes to me and says that they have a salesperson that is a pretty good salesperson, but needs help closing, I immediately know that there is probably nothing wrong with their closing techniques or style, but there is probably something wrong with the whole process.

The reason that tricks, techniques, and fancy closing questions just don't work in our business, is that whether you're installing telephone equipment, connecting telecommunication services, or selling local or long distance services, you are initiating, and entering into what should be a long-term relationship with your client.

A dear old friend of mine, a prominent attorney from Hartford, Connecticut, who sadly passed away in 1996, Joe Fazzano, used this wonderful analogy about relationships.

THE POSTAL WORKER ANALOGY

When Joe Fazzano was engaged to his wife, they discussed the kind of relationship they wanted. He promised to treat her like he would want to treat the postman (or should I say postal worker) that has been delivering mail for the last 10 years, and probably will be delivering the mail for the next 10 years. If you saw that postal worker drop a couple of your letters on the muddy ground either by accident or because of what you may have perceived as carelessness, what would you do? You wouldn't go out there and chew the person out. If you did you might not see another "on time" delivery for a long time. Likewise you wouldn't react negatively to similar actions of a new spouse, a spouse that you were about to spend the rest of your life with. Likewise, in the Telecommunications Industry, we are building relationships with our prospects and customers that will last for many years —relationships that are ongoing. There are no short term tactics that can be or should be used here.

Longterm relationships take time to develop.

You could have yelled out sarcastically to the postal worker..."Hey stupid...you dropped something," and the postal worker might have apologized. But, I can promise you that you would pay a dear price for a long, long time. Treat your prospects with the same respect during the closing process that you would treat anyone that you foresee a long-term relationship with.

THE SALES PROCESS

"Closing" in the Telecommunications Industry should be done naturally, with lot's of empathy, and above all should be a logical conclusion of everything else done right during the sales process.

This chapter on closing represents the seventh out of eight steps in the proven effective sales process to success in selling telecommunications stuff. I call it a process, because just like any other process, if you skip a step in a logical process, the results could be suspect. You might not even get the desired results at all. Just ask any brewmaster, involved in the making of beer. How about talking to the owner of a winery and asking about the process of making fine wine. Ask a major league baseball manager about the development process of a young ball player, or ask a marathon runner about the process or preparation necessary to run a 26.2 mile marathon, and the steps that are necessary for success.

Just like in brewing beer, making fine wine, and developing a major league ball player, or running a marathon, selling in the Telecommunications Industry is a process. Skip a step, fail to master the process, and you will fail. The beer will taste bitter, the wine won't be smooth to the palate, slightly fruity, and full-bodied, the ball player will be unable to hit a major league curve ball, the marathon runner will hit the wall at 22 miles, and most importantly, you won't close the sale.

Before we review the key aspects of the selling process, the process that puts us in a position to ask a prospect to spend hundreds, thousands, or even millions of dollars of their company's money, I'd like to say a word or two about closing.

If you don't close, who loses? Well, a lot of people lose. You lose, because you don't get paid a juicy commission check at the end of the month or a bonus check at the end of the quarter.

Your sales manager loses, because he or she doesn't get paid the extra they get paid for their supervision, motivation, and support.

Your family loses, because they don't get to enjoy the comforts that the extra money will afford them.

Your branch manager loses, the regional manager loses, the VPs lose, the president loses, the board of directors lose, and the shareholders, if any, lose.

Most importantly, you must believe in your heart of hearts — the prospect loses. The prospect loses because they don't get to enjoy all of the benefits they can enjoy from your product or service.

There certainly is one more person that loses — I lose. I lose because I take this business very seriously. The only way that I prosper is by helping you. That doesn't mean just helping you learn a little more about selling, or helping you learn some interesting things about selling telecom stuff, but helping you enjoy the process of selling a little more. In the long run, the only way it makes sense for me to write a book like this is if I help you sell more. In the final analysis, that's the only reason that you bought this book. Unless you believe that reading this book will allow you to sell more, or enthusiastically go to your manager and tell them that you sold more because of this book, this whole process isn't working. The only way you sell more is if you "CLOSE" more.

REVIEW

So let's review briefly the logical process of selling in the telecom industry, which really is the process of closing in the telecom industry, the process that has been the subject of the previous chapters.

The objective of Chapter 1 was to ensure that you feel great about the profession of selling in general, and are truly excited about selling in the Telecommunications Industry.

In Chapter 2 we laid out the eight-step sales process for selling in the telecom industry that will ensure your success. You were reminded that the process is just that. If you skip any step, then the next step becomes moot.

Chapter 3 was all about the first step in the sales process — developing the necessary prospecting and networking skills to find qualified prospects for your equipment and services.

Without qualified prospects, and without a powerful "initial benefit statement" or "value statement" prepared to encourage them to meet with you, there was no use in reading Chapter 4. That was because Chapter 4 outlined exactly what to do during that first appointment (Step 2), and you were coached during that very first appointment with the prospect to "open up" the close.

You can't close the door, unless you've opened it first. To further test the waters, I suggested strongly that during the first appointment, you try a trial close. An outline of a typical trial close is something like: "If I could do such and such, would you be in a position to do such and such?" If you did it, then at this, the closing stage, your chances of closing are dramatically increased. If not, then who knows? We also covered in Chapter 4, some key objectives for the first appointment, such as:

- How to establish rapport.

- How to gain trust and confidence in yourself, your company, and your product.

- How to be totally in control.

- How to position yourself for the long term.

- The basics of qualifying.

- The importance of getting a commitment for the next step.

Because qualifying is so important in our industry, or disqualifying, as we came to refer to it in Step 3, I dedicated an entire chapter, (Chapter 5) just to qualifying:

- Qualifying for Money (Do they have it...have they budgeted some?).

- Qualifying for authority (Who will make the final decision?).

- Qualifying for need (Do they just want it, or really need it?).

- Qualifying for an impending event (A move, a change, or big existing problems?).

- Qualifying for their application (What are they going to do with our stuff?).

- Qualifying for competition (Who else are they looking at and why? Who is the incumbent? Who is favored?).

- Qualifying for a time frame (Where we go from here?).

If you qualified well, then the closing step in the sales process will be so much easier.

In Chapter 6 (Step 4) we began a thorough fact-finding stage. In that chapter you were reminded that "knowing something about your prospect is as important as knowing everything about your product." At the end of that chapter, we looked at the variables that must be present before your prospect buys. What this means is that one or more of those variables must be present before you can close, and if they're not present, you can ask all the closing questions you want, you can stay silent for hours, but unless one or more is present you won't close.

You potential customer must have a desire to:

- improve profitability through telecommunications
- increase revenue with telecom equipment or services
- reduce expenses directly or indirectly by buying telecom stuff
- improve productivity of their people through telecom stuff
- reduce inefficiency
- save money
- save time
- increase market share
- improve reliability
- improve the safety or security of their employees

The more specific information you have about how the prospect is going to (1) satisfy their desires with your stuff and (2) cost justify the expense; along with the knowledge of how your products and services do it better than your competition, the better your chances are of closing.

By the time you got to Chapter 7 (Step 5 - Effective Presentations), you were really into thinking seriously about closing. In that chapter we covered a detailed step-by-step process within a process for ensuring a high percentage closing ratio — if certain things were done prior to inviting the prospect into your office for a presentation or demonstration.

Prospects have questions, prospects have concerns, prospects have a strong desire not to make a bad buying decision, prospects are influenced by the competition, and, most importantly, prospects prefer to tell you little white lies rather than the real reason why they won't buy from you. For these and other reasons, Chapter 8 (Step 6) was dedicated exclusively to overcoming objections.

You learned the five different types of objections and how to handle each professionally. You learned how to find the REAL objection. You were taught, and now are proficient at the proven-effective seven-step process for handling objections. You know how to anticipate objections and how to actually bring them up during the sales process. You learned to be prepared with sales tools such as:

- testimonial letters

- product comparison charts
- market share documentation
- national account user lists
- reference lists
- financial analysis

These tools help you overcome traditional objections in our industry and move toward a close. Most importantly, you now know that objections are a natural progression toward the close, and without them you probably don't have a qualified prospect.

So with all of that background covered, let's talk about closing.

CLOSING TOOLS

Here are 3 important things I want you to remember about closing:

1. Closing is a natural outgrowth of doing everything else right during the sales cycle.

2. You must close with warmth, with empathy, and with lots of feeling.

3. Think of closing as the culmination of what all of selling is really all about — Helping people make decisions that are good for them and their companies.

   How do you know when it's time to Close?

   - It's time to close when the prospect slows down the pace.

   - It's time to close when there just doesn't seem anything more to cover.

THE TRANSITION

Over the years, I have heard numerous salespeople express concern about their closing. Usually their concern, however, is not actually in closing, but in knowing how to transition into a closing question. They expressed the concern, and rightly so, that an ill-timed closing question could alienate their prospective customer. One of my favorite stories, related to me by my dear friend from Newport Beach, California, Dan McBride, is as follows:

Dan walks into an automobile dealership and the salesperson walks out, looks Dan right in the eye, and says, "What's it going to take to get you into an automobile today?" Dan looks him back right in the eye, and says, "another salesperson!"

Obviously, an ill-timed closing question.

A poorly timed closing question can have a devastating effect on the sales process.

So how do you get the nerve up to close? You might use a soft transition comment and say something like: "Well that just about covers everything." Or "We seem to be in harmony on all of the major issues." Or "This really is a good fit." Then you move toward a closing question.

There are lots of different types of closes or closing-type questions. Here is a list of my favorite six categories of telecom closing statements. Almost every time I've done a sales training seminar, I'll ask the salespeople their favorite closing question. Invariably I'll find someone that will insist to the class that their closing question is the best. What this tells me is that if you find one of the following that works for you, you can use it over and over again. Quite frankly, I think you should have a number that you like, and use the appropriate one for different situations.

CATEGORIES OF CLOSING QUESTIONS

1. Alternative of choice closes for the telecom industry

2. Assumptive closes for the telecom industry

3. Summary closes for the telecom industry

4. Extra-incentive closes for the telecom industry

5. Conditional closes in the telecom industry

6. Direct telecom industry closes

So here are my closing examples, each customized for the Telecommunications Industry.

Alternative Closes

Will we be training the executives in a separate group, or will they be trained along with everyone else?

Will we be able to cutover your new system during working hours or will we have to do that on an off-peak hour?

Will you be going ahead with the basic variable-rate calling pattern program or the customized fixed rate calling pattern program?

May I use the corner of your desk to write up this telephone order, or would you prefer I set up over there?

Assumptive Closes

We seem to be in total agreement with this configuration, is now a good time to complete the paperwork?

Well that seems to cover just about everything, with your approval we can get you enjoying the cost-savings next month...May I fax you the paperwork for approval?

Well, that settles that, the only way to solve your major telecommunications need is to go ahead and complete the paperwork, is now a good time to get started?

What a nice fit for your application...should I send the paperwork regular mail, or next day delivery?

Summary Closes

So John, you can see the substantial cost-savings that will accrue to your company with this calling plan...when would you like to put our proposal into operation?

So Susan, since we agree that you're comfortable with our service and support, and the system meets your telecom needs, and we're price competitive, then the only thing left is to get your approval on my agreement. Would you endorse the contract right here?

Since we agree that time was your major concern and since I've satisfied your requirements with regard to service, design and price, may I call the office to set up an installation date?

So can I summarize for you the major benefits of going with us? First of all you get the lowest priced calling plan in the industry, secondly, it's backed by the service of a major telecom company, and thirdly, our guaranteed quality, or your money back...no questions asked...doesn't that cover everything?

Extra Incentive Closes

If you order now, I am authorized to include an extra second year warranty at no charge...that's over a $500 savings.

We are having a special promotion this month. If you sign up during February, you will get an extra 10% discount. Don't you think that's enough of an incentive to go ahead now?

Our company would like to have you as a customer for a long time...I am authorized this month to offer you as a new customer an extra 10% discount. Doesn't that make your decision to go with us a little easier?

During this promotion, we will include the installation of your system free of charge. Aren't you pleased?

Conditional Closes

If I were able to get you credit approval at just two points over prime, would you be in a position to sign our agreement today?

If I can take care of the price differential between them and us, would you be able to complete the paperwork today?

If I can arrange to get you included in the special introductory offer category, could we go ahead and complete the paperwork?

If I can get approval to have your system installed by the end of the month, are you in a position to go ahead now?

Direct Closes

What will be the purchase order for this requisition?

Do we have an agreement?

Does this agreement meet with your approval?

May I use the corner of your desk to write this up now?

So there you have many of my favorite telecom closes.

If you look them over, you'll see that it is easy to mix and match phrases from each. I'd encourage you to mix and match the phrases, and find words that you're really comfortable with using. Over the years, I've found that my closes came pretty naturally as an outgrowth of all the work that went into getting into a position where the close was easy.

I do have to emphasize, however, that you do have to ask, and when you ask, it should be well thought out so your closing question is strong. Remember this though, after you ask a closing question: "The first person that speaks loses"…"The first person that speaks loses"…"The first person that speaks loses."

SALES MANAGER'S EXERCISE:
Ask your salespeople for their favorite closes. If they don't have one, that should tell you something. Review the list of my favorite closes in the industry. Have each salesperson pick one they like or a combination of a few they like; role-play using the closes.

CUTTING THE PRICE

One large client of mine, a large telecommunications manufacturer, recently introduced a program authorizing the salespeople to take an extra percent discount off the list price during a promotion they were running. This is pretty common in the industry, and lots of salespeople do it. You can see how easy it would be to say: "If I can do something about the price differential between their price and ours, will you be in a position to go ahead today?

Unfortunately, when that's done, you as the salesperson lose a little of your commission, and the manufacturer and the distributor lose a little of their margin. It's the margin, the difference between the loaded cost of the product and what prospects are willing to pay, that is the fuel that keeps us all charging around in the sometimes crazy world of telecommunications. If we lose our margins, we'll all go away. You'll lose your job and I'll lose my job training you.

You and I both have a vested interest in trying to sell our value to the customer at list price. If you are faced with a competitive price lower than yours, try one or more of the following **before** cutting your price.

Ten Things to do Before Cutting Your Price

1. Hold your ground - Don't just lower the price for the sake of lowering the price. There is no easier way to lose respect in the eyes of your prospect. Remember, many of the prospects you are working with came up through the sales side of the business. They know what selling is all about, they might just respect you more, not less, if you hold your ground.

2. Think profitability - You can remind them that they have to sell their products and services at a profit in order to stay in business. Your company also needs to sell products at a profit to stay in business.

3. Think quality - You might ask them if they've ever heard the expression: "The bitterness of low quality remains long after the sweetness of low price is forgotten."

4. Equate your price to what they're getting - You might point out that: "You get what you pay for in the Telecommunications Industry." This is a risk/reward business. The lower the price, the higher the risk of poor quality, shoddy installation, poor documentation, unacceptable response times, poor service, all resulting in traumatic effects on your company's image and competitiveness!

5. Don't fall for this - Be aware, that negotiating 101 teaches that when quoted a price of anything...you should respond with astonishment. You can tell you're up against a real pro when you show them your price and you say ... "Your investment is $17,000 dollars"...and they respond in astonishment: "$17,000 dollars!" They are just waiting for you to say; "Well if that's a problem...maybe I can work on the price a little." By just feeding back to you your price, they've just saved a few thousand dollars, you've just lost part of your commission, and our whole industry just got weaker.

The correct response is: "Yes, $17,000 dollars, we talked about that price range earlier, or if that's a problem, we can look at taking out some of the extras you wanted, or revising the configuration, or don't you think we're worth it?"

6. Think value not price - Before you cut your price, ask the customer to reconsider the value of doing business with your company — go back and review what he is getting — the service, the support, the local or worldwide presence, the quality, the no-nonsense guarantee, the professional installation, the warranty, the guaranteed add-on cost of equipment, etc.

7. Make sure they are being straight with you - Before you cut your price, make sure that the customer is being straight with you. Sometimes they will tell you they have a lower price, even if they want to go with you. If you hold your ground they will go ahead with your original price.

8. Check the competitive quote - Check to see if the competition has included everything in their quote. Maybe they made a mistake, or maybe they left something out on purpose just to get the order. Ask for their configuration in order to make sure you are comparing apples to apples.

9. Relate the potential horror of going with the lowest price - Before you cut your price, give the customer an example of horror stories of other customers that went with that competitor and got what they deserved. Sell the value of going with you at the higher price.

10. Relate examples of quality, price, and satisfaction - Give the customer examples of industries where there is a direct relationship between price, quality, and satisfaction. Certainly this is true of the auto industry. First, make sure you ask them what kind of car they drive, then let them compare the difference between a Chevy and an Oldsmobile, or whatever. Think about cuts of beef. Have you ever ordered the cheap steak? Have you ever paid a little more for a higher cut of beef, only to have the guests rave about the food. Wasn't it worth it? If I had to ask you, you could probably name the three or four most expensive pieces of clothing in your closet. Aren't those expensive clothes probably the clothing of choice when you really have to look and feel good?

I don't want anyone to think for a minute that I am naive about price negotiations. In many industries, many business people just won't buy anything unless they get a price break or at least think they are getting a price break. So if all ten of the approaches listed above fail; then make sure you know exactly what the price differential is, who you are up against, and offer to meet the customer half way.

MOTIVATIONAL SECTION
I'd like to end by sharing with you a recent study at Harvard Business School.

They did a study to determine the common characteristics of the top salespeople in the country. Here are the eight most important attributes of the top sellers:

As you read each one, why don't you rate yourself on a scale of 1 to 10, with 1 being a real weakness of yours and 10 being an absolute strength. You'll get some insight into some character traits that you might need to work on. I'll share with you my rating and why:

1. They do not take no personally.

They do not allow it to make them feel like a failure. They have high enough levels of confidence and self esteem so that although they may be disappointed, they are not devastated.

I filmed a whole video on trust and confidence and know a lot about the importance of feeling confident. I've skied some of the steepest ski slopes in the world; like the Sudan Coulour at Blackcomb, and Corbets Coulour at Jackson Hole, WY. I've lined up in Hopkinton, MA. (the start of the Boston Marathon), 10 times. I dove to over 100 feet in Cozumel and the Grand Cayman. All with no sweat, no fear, total confidence...but when someone says no to me, I do take it personally. I have to admit, I do get down. I'm not really devastated, but sometimes I can't shake it. I hold it inside far too long. Eventually, I do go on. I guess it just takes practice, the ability to go right through it, pick yourself up, dust yourself off, start all over again. I do hate to lose. I give myself a 6.

2. They take 100% responsibility for their results.

They don't blame the economy, their company, their pricing plan, or their territory for lack of results. The tougher things got the harder they work.

Hey, I run my own business. If I don't get excited about writing this book, who will? If the book doesn't sell well, who's to blame? If I can't pay my mortgage, who will? You wouldn't be in sales, or shouldn't be in sales, if you can't take responsibility for your own success. There is only one person responsible for your success, and that is you. I give myself a 9.

3. They have above average ambition and a strong desire to succeed.

This is pretty key, because it affects their priorities, and how they spend their time on and off the job, and with whom they associate.

I've always liked to work hard. But I've always also wanted to play hard. I'm not and never will be a workaholic. I've never wanted to be real famous or real rich or whatever. The quality of my life has always been important to me, my relationship with my daughter and loved ones is more important that another sale or a promotion. That doesn't mean I haven't been focused, I just usually try to find a balance in life between work, play, and family. I give myself a 7.

4. They exhibit a high level of empathy.

The ability to put themselves in the client's shoes, feel their needs and desires and respond accordingly is relatively easy for these top sellers.

I speak and train salespeople all over the world. I ask my clients to have their salespeople take multiple days out of the field, that costs them money and selling time. I ask them to pay me a fee, and all my expenses. If you're not good, if you don't show empathy for their expense, their cost of doing the training, they probably won't spend the money. Think for a minute what you're asking your clients to do. Empathy is a big, big issue. I give myself a 9.

5. They are intensely goal oriented.

They seem to always knowing what they are going after and how much progress they are making. They keep distractions to a minimum and didn't get side-tracked.

Not my cup of tea. I've spent a lot of my life not knowing exactly what I wanted, or even knowing what I was good at. I sold, I was in marketing, I was a consultant, I sold again, I worked for a big company, I worked for a small company, and along the way I had and continue to have a lot, and I repeat, a lot of fun. I get distracted easily. A tarot card reader once said it's the joker in me. Other's would say it's my Gemini personality. If you can't have fun doing what you do, get out of the business, you'd be a bore anyway. This is hard to admit as a sales trainer, but I've gone for months without written goals. I sort of wing it a little. I do, at least, always have a pretty good idea of what I should be doing, and don't get too far off track. I give myself a 5.

6. They have above average will power and determination.

No matter how tempted they are to give up, they persist toward goals. Self discipline is the key.

I don't have a problem with persistency. When I'm selling and I find someone that I know has to make a decision, I am one persistent person. Discipline is an issue for me. I need to acquire more of it — discipline to write out my goals, discipline to plan ahead better, discipline to make the telemarketing calls, discipline to update my database, discipline to follow-up diligently with customers, to sit down and finish writing this book. I give myself a 7.

7. They are impeccably honest with themselves and their customers.

No matter what the temptation to fudge, these people resist and gain ongoing trust and confidence of the prospects.

I learned this early. I'm a 10.

8. They have the ability to approach strangers even when it's uncomfortable.

As a salesperson, you need to practice this. Next time you're in a line, try this. See if you can strike up a conversation with the person ahead of you, or behind you. Just out of the blue, see how much you can find out about them, see if you can get them to smile, see if you can get them to like you.

An old friend of mine, Eddie Samp, met his wife while waiting in line to vote in a national election. He told me that as he walked up to the line she was last in line. Eddie addressed her politely, "Excuse me, are you last in line?" She looked around and replied, "Yes I am." Eddie simply smiled and said, "Not any more, aren't you glad I came along?" Eddie is still one of the finest and most successful salespeople I know.

I've never had trouble meeting and engaging people in conversation. Give me a 10.

Step 8:Follow-Up Support

So you asked a closing question and your prospect said, "OK". You've arranged for the financing, or you have the down payment in you briefcase. You're on your way home to your office and you think you've done your job.

That's wishful thinking — you're job is not done. I love the expression: "The laundry's not finished till it's folded and put away."

The laundry's not done till it's folded and put away.
The sale's not over until you've got a valued reference.

There is more work to be done, but before you continue with this, the eighth and last step in the sales process, it's time to celebrate. I'm not kidding. Before you continue on and do the work of the final step in the sales process — the follow-up and support — it is important to take a moment and reflect on what just happened. Someone just bought something from you, they've given you money, or promised to give you money for your products and services, and they expect to get what they just ordered.

When I ask salespeople in my seminars, what they are selling, most will respond that they are "selling themselves," and that's correct. Most of the time the buyers are "buying" the salesperson. That doesn't mean that they're not interested in the products and services you offer, but they are relying on you to translate their needs back to your company. Earlier in this book, you learned that "knowing something about your prospect is more important than knowing everything about your products and services." Now, after they've bought from you, it's time for you to translate their needs back to the people in your company that can fulfill those needs.

Most companies have a job package, or job sheet that you'll need to fill out. Sometimes this is simple, sometimes this is extremely time consuming. That's why I say: "Before you start this last step in the sales cycle, it's time to celebrate." Take a breather and rejoice in what has just happened to you. Someone has just expressed confidence in you. Someone has just bought your ability to take care of them. Because we're talking telecom, I know that what you've just sold is mission critical to the overall health of their business. This was not just a trivial sale. We're talking about stuff that can impact the total way they do business, have an impact on their revenues, their profits. You can impact their shareholders, their net worth, their well being, even how well they'll sleep at night.

So take a moment, or two, and celebrate their confidence in you. They have just bought you, and are about to pay the price for your abilities.

When I finished college and done my tour in the Navy, I went looking for my first sales job, and my father would often give me advice about the profession of selling. One thing he would always say is, "it doesn't matter if it is a big sale or a little sale, the moment you stop getting a real kick out of someone giving you a check for your products or services, it's time to get out of the business, remember, they've just bought you."

Get a kick out of getting the check, or get out of the business.

Those are good words of advice, and I've passed them on many times. Selling is not easy, and when you make the sale, it is time to take a moment and reflect on what just happened, especially where there was tough competition involved. You have just won, you have just beaten the competition, and you have just been given a vote of confidence. So take a moment or an evening and celebrate their vote of confidence in you. Take a moment to celebrate your success, and enjoy the moment.

STAY INVOLVED

In many industries, the sale is the end of the sales process, but it is not so in telecommunications. From the very beginning of this book we set up the suggestion that the sale was based on establishing a long-term relationship. Some companies I've worked with position their salespeople to make the sale then deliver the client over to those that can implement the sale. Others prefer to have the account executives stay involved. I'm going to suggest, even if you are totally responsible only for new sales, that you stay involved. The degree to which you stay involved will depend upon a number of variables, such as the complexity of the sale, the size of the sale, your implementation team, and the sales and commission structure of your company.

Have you ever bought something, and it has not lived up to your expectations?

All of us can think of small or large purchases that disappointed us. Sometimes it is impossible to point to just one characteristic of what you bought, but when put together, it all didn't add up to what you thought you were buying. On the other hand, can you think of a product or service that exceeded your expectation. Remember how pleased you were. Here's what I want you to think about, mostly during this follow-up and support step in the sales cycle. Try to do everything in your power to ensure that your company delivers more than expected to your new customer. Try to do the little things to make your customer feel that they got their full moneys worth, that they got full value for their purchase.

In short, your new customer is relying on you to perform and will, to varying degrees, hold you responsible for making sure that your company performs. Here's the key point — to the extent that you and your involvement can ensure that your company performs satisfactorily — you will be able to use your new customer as a valued reference.

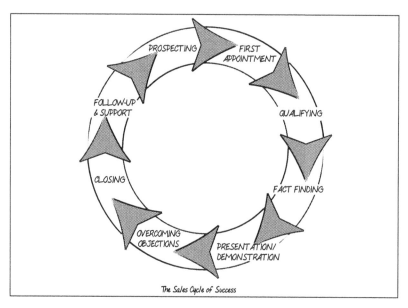

The Sales Cycle of Success

If you look at the above illustration, you'll see that the eight-step process becomes a continuous process. If you do your job correctly, the successful sale and implementation of your products and services will lead to a happy customer, and a valued reference. Which flows directly back to Step 1, and allows you to find more qualified prospects, gives you a competitive advantage during the first appointment, and makes qualifying a lot easier.

REFERENCES

In Chapter 3 on prospecting and networking we discussed the importance of reference selling. The sales experts will tell you that strong references and testimonials are the best way to convince someone to do something.

If this is the case, then it's obvious to see the importance of developing a strong reference list. If you're a sales veteran, then all I want to do is reinforce the importance to you of continuing to develop strong personal and professional references. If you're a rookie, it's time to set some goals.

Set a goal of developing five fabulous references. The time frame for this is really up to you, but I'd suggest that you give yourself a year to accomplish this. The kind of references I'm looking for are those where the customer has had direct contact with you and can explain to a prospective customer why it is worth their while to do business with you.

Remember way back to one of the very first stories I shared with you in this book — the story about Allen's pharmacy. Jay bought his new phone system because of what he had learned from the salesperson. To Jay, even though two salespeople were proposing the same system, the differentiating factor was the salesperson. The best way to convince anyone of your value is to have that person call a customer reference.

Only after you've sold someone something, then fully delivered or ensured that your company fully delivered on what you promised to the customer, can you begin to develop the kind of relationship that you can use as a reference. Once developed, these references will become your best closing asset. Solid references and testimonials are priceless when it comes to closing.

REFERRALS

Referrals are what really make our 8-step sales process a continual process. The long-term development of references for referrals is probably what most separates the veterans from the rookies in the Telecommunications Industry.

The veteran salespeople that are the most successful have developed a number of personal references who not only speak highly of them, but who would not hesitate to refer a friend or business associate of theirs to them.

When you are highly recommended to someone, and walk in for your first meeting with this individual, the whole environment is different. There is immediately an area of common ground. You and the prospect can share your experiences with the person that referred you. If the referral is set up right, this should be an upbeat conversation. When I ask someone for a referral I always ask about their relationship with the person they are referring me to. I like to know as much as possible about the details of both the business relationship and the personal relationship. That way I immediately have material to use to break the ice with the new prospect. In addition, it is important to ask your reference, if they'll call your new prospect, and introduce you.

If you do these two things, get details of the personal and professional relationship, and have your reference call your new prospect before you meet, then you will have a huge competitive advantage early in the sales process.

Initial meetings usually have tension. Sometimes it is overt, and sometimes the tension is subtle. The prospect knows that you're there to try to sell something. You're following the sales process I've laid out in previous chapters. Specifically, you're trying to remember some or all of the ten objectives for a first appointment (Chapter 4).

If you're first appointment with a prospect is from a referral, think how much easier it will be for you to meet your objectives. You come to the meeting highly recommended, which means there is an implicit level of trust. You come to the meeting with a high level of confidence, which means it will be much easier to gain the confidence of the prospect. You come to the meeting with some common ground to talk about, which means it will be easier to transition into other areas of commonalties. Because you and your company come highly recommended, it should be easier to establish credibility in your company, its products, and services, and to establish a partnership relationship.

Coming to the first meeting highly recommended, means you can focus more of your energy on being well prepared, qualifying, positioning yourself for the long term, setting a time frame, and being in control. You're way ahead of the game, and well into the sales cycle of success.

MOTIVATIONAL SECTION

I'd like to share the experience of a highly successful salesperson that not only follows the eight-step sales process you've just learned, but who is particularly diligent about the last and final step of follow-up and support for valued referrals.

Peggy Nelson sells major telecommunication systems for Exp@Nets, previously Lucent Technology. She has over 11 years selling experience, almost 5 years with AT&T, and previously 6 years with Southern New England Telephone Company (SNET). She has consistently been a high achiever, consistently in the top 5. Recently she was third out of over 50 sales reps.

Peggy has won all kinds of awards, including cash awards, a cruise to the Bahamas, a trip to California, and a trip to Florida. The award she is most proud of, however, was the Summit award. This award is not only given for meeting and exceeding revenue goals, but also for being judged #1 by her customers, peers, and coach. Peggy readily admits that following the process and putting the focus on her customer's satisfaction is a big part of her success.

When I last saw Peggy, she was driving around in a red, BMW Z3 roadster convertible. Guess what? She won it for being over 150% of quota for the first quarter. Now Peggy has won a lot of stuff for her achievements in sales, and while she acknowledges that she has been recognized often for her accomplishments, nothing has meant more to her than winning that car.

For the first time in her life, Peggy really feels successful. Everywhere she goes in her flashy car, people notice her. Her family and friends, her neighbors, and especially her customers ask about the car; and she gets to explain that she won it for being one of her company's top salespeople. The car has made her feel more positive about her company, more confident and more successful. It is true that success breeds success.

Here are two final thoughts about Peggy and her red convertible BMW Z3 convertible. First, of all the things that Lucent and now EXP@nets could have done for their top performers, they found that they are getting a lot of bang for their buck by awarding these automobiles. You could feel the excitement in Peggy's voice and her enthusiasm for her profession is heightened each day as she gets into her convertible with the top down and the heated seats turned on.

Second, a thought about the state of the telecom industry. With the industry characterized by mergers, acquisitions, consolidation, and growth, even if a salesperson stays with the same company, they can find themselves working for numerous companies. Peggy's experience of SNET, then AT&T, then Lucent, then Exp@nets, is typical. I've worked with thousands of sales reps — one never changed companies, but worked for New England Telephone, NYNEX, Nortel

Meridian Systems, Wiltel, and Williams Communications Systems, and another worked for Executone, Executone Business Solutions, Claricom, and now Staples Communications Systems.

In this era of constant change, managers need to do everything possible to continually make their salespeople feel good about themselves, and especially about their company. Peggy sure feels good about herself and her company.

Success and Failure

So there you have it, my absolute best thinking on "How to Compete and Win in the Telecommunications Industry." I've (hopefully) convinced you that telecommunications is the place to be at this moment in time because:

- It is a rapidly growing, evolving industry — an industry characterized by favorable regulatory, legislative and judicial rulings.

- The rapidly changing technological trends create all kinds of career and especially sales career opportunities.

- The convergence of computers and communications, and the proliferation of companies in the local, long distance, wireless, equipment, networking, and Internet areas make it an exciting arena to work in.

- I've shown you the secret to selling success in the Telecommunications Industry through my proven-effective eight-step sales process.

With regard to PROSPECTING, you learned the best types of prospecting, and the factors that would prompt someone to want a new telecommunication system or services, but because most salespeople hate prospecting, you've learned the best way to avoid it altogether and still be successful.

With regard to your FIRST APPOINTMENT, you learned the danger of applying generic sales training techniques to the principles of selling in the telecom industry. You know that decisions are based solely on product, price, and service, and that you must position yourself for the long term. Any attempt to qualify too heavily, attempt to manipulate the sale, or close too early will be met with resistance and failure.

If the initial meeting has gone well, you moved on to QUALIFYING, and you learned all about the importance of asking great questions, but specifically the best questions to ask to uncover the real telecommunication requirements.

You positioned yourself and were given permission by your prospect to really learn about their business or organization. Now that you're FACT-FINDING, you learned the process for uncovering their real telecommunications needs. More specifically, you learned how to ask questions about the incoming flow of all types of communications traffic such as voice, data, image, fax, video, and Internet. And then to move on to questions about the outgoing flow of communications traffic. Then, if appropriate, to ask questions about internal traffic.

When it came time to present your findings, or solutions to your customer, and to give an EFFECTIVE PRESENTATION, you knew what you needed to complete during the fact-finding stage to be prepared for your presentation. Armed with a thorough understanding of your customer's telecommunications areas of concern and their telecom objectives, you learned a process that ensures over a 90% closing ratio during the presentation.

As you neared the end of the sales process, you read about the top 5 most commonly heard objections in the Telecommunications Industry, and the top 20 real objections in the Telecommunications Industry, and learned a 7-step formula for OVERCOMING OBJECTIONS.

Now you know that you should welcome and not fear objections, because if you handle objections effectively; you'll be prepared to CLOSE. You now know what to do during the entire sales process to ensure a "natural" close, you know when it's time to close the sale, and you know the best closes in the Telecommunications Industry.

Lastly, you know that after you close, your job is not done. It's time to FOLLOW-UP AND SUPPORT your customer to ensure that you deliver to them more than they expected. It's time to maintain the relationship in order to gain a valued reference and a strong referral to another prospective customer, so you can continue the sales cycle of success.

If you practice the process you will win, and if you continually win you will be successful. Because the word "success" keeps coming up, I'd like to conclude by sharing with you three success stories, and finish with a personal failure story.

Think of *success* as the "journey", not the "destination." This mind set will save you a lot of heartache along the way. Try to avoid a thought process, such as: When I make VP, I'll consider myself successful, or when I get a corner office, I'll be successful, or when I make a million dollars, I'll be successful, or if I lose this deal, I'm a failure.

Try to think of success as the process, not the end result. The following stories illustrate the importance of this type of thinking.

THREE RUNNING BUDDIES IN EQUADOR

Recently, three running friends of mine decided to fly to Equador to climb the second highest mountain in Equador — Mount Cotopaxi. Now these three friends, Rick Silverman, John Boyle, and Rich LeBeouf are not your ordinary couch potatoes. All three are highly trained, well conditioned athletes. In fact, Rick and John have each run over twenty-five 26.2 mile marathons, most in under 3 hours, and Rich, the third member of the party, has run at least 15. In fact, over the last few years, the three have stretched their training to include 100-mile runs; Rick having completed 4, John 3 and Rich 2. The three climb, run, bicycle, and work out regularly.

All three, enthusiastically prepared for the mountain climb. They realized that it certainly would be a challenge to climb the 19,700 foot peak, but felt confident that since they were in top physical condition, they could reasonably expect to realize their goal. Why would they want to climb this mountain? Because it's there. How would they measure success? Getting to the top. Isn't that what mountain climbing is all about? Isn't that what you do when you "climb" a mountain?

They flew to Quito at 10,000 feet and spent a few days getting acclimatized or adjusted to the high altitude. Then they were driven up to the 14,000 foot level. Just walking around at 14,000 feet is an effort. On day one they climbed to the "refugio," a small shelter high up on the mountain at 15,500 feet. Day two began at 12:15 AM, pitch dark, with brilliant stars, and the summit at 19,700 feet their goal. "Cotopaxi or bust" was ringing in their ears.

To hear Rick explain it, what happened next was one of those experiences that alters the rest of your life. After over 7 hours of climbing, at about 7:30 in the morning and at 19,500 feet, just a mere 200 feet short of their goal, a goal that they had worked toward for many months, Rich LeBeouf got sick. He was suffering from high altitude sickness, dizzy, tired, and slightly nauseated. He couldn't go another step forward.

Rich was in bad shape, he had to lie down. As Rick and John laid him down their eyes met, and without saying a word, they knew they only had a couple of choices. They could both leave Rich, and return a couple of hours later, or one or the other could push on to the summit, and return. Those thoughts were only fleeting as both immediately began to realize the severity of Rich's ailment....they had to get him down, and get him down to a lower elevation as soon as possible.

In a matter of moments, their definition of success had changed. They looked at each other, and knew what they had to do. In many ways, the decision to forego the physical challenge of making it to the top, and make the mental adjustment of taking care of a friend, was a higher level of success. This was not failure, this was part of life's experiences, this was part of the journey.

By the way, they got Rich down safely, he recovered rapidly, they got home in good shape, all three still like and respect each other, work out together, and dream of the next adventure.

MICHAEL CRISTALDI, THE BARBER

Today, they would call him a hair stylist, I prefer to call him the man that has been cutting my hair for over 20 years. I like to joke with him that he should only charge me half price, and you know why if you look at the author's picture.

Michael came to this country from Italy in 1957. He left in the middle of high school when he was 15 years old. He has been cutting hair, 5 days a week, since 1963. From the very first time Michael cut my hair, it was obvious that Michael loved his job.

"This is not like work to me," he said as I asked him the questions for this book. He moved into his prestigious Coply Plaza hotel location in downtown Boston in 1976, and with pride he told me he cuts the hair of John Havlichek, the famous Boston Celtic basketball player of the 1970's, previous mayor of Boston, Kevin White, previous Governor of Massachusetts, Gov. King, and others.

Early on I realized there was a lot more to Michael than what you'd think. Every now and then, while in the middle of a haircut, he'd apologize and take a business call. While I could only hear one half of the conversation, it was about a lot more than cutting hair. You see, Michael has amassed a pretty good size real estate portfolio. He told me that he bought his first two-family house in Roxbury, a part of Boston, over 20 years ago. He put $14,000 down on this $70,000 piece of property. Today he owns and manages 40 apartments in 8 to 10 different buildings. The cash flow is over $500,000 a year.

When asked the key to success in getting to this point in real estate, Michael would say that it has been a lot of hard work, nights and weekends. For many years, he managed, and did most of the upkeep work himself. In recent years, he has hired and watched closely the best contractors he could find. He keeps the buildings in great shape, responds timely to tenant problems, and charges below market rates in order to keep costly turnover to a minimum. Sounds like the kind of landlord I would want.

A few years ago, when he realized his wife would have to stay home with his new son, he opened a day care center. This has gone so well that soon he will open a second day care center.

Hair stylist, day care center owner, real estate holdings, fine young son and wife, prestigious clients, sounds like success to me; and it is, but Michael has never felt like he has arrived. He still is on a journey. Recently, we chatted about the Internet; about Home Depot, Nokia, Qualcomm, and Sapient stock offerings; about wireless communications; and about quality educational opportunities for his son. Michael does stuff. Many times, while chatting with him, I've thought that he must be tired when he gets home after cutting hair all day. I'm sure he is, but that hasn't stopped him from pursuing his other financial interests. He has quietly and relentlessly built a solid financial foundation for himself and his loved ones. Not your typical Barber, just a man on a journey, a very, very, successful one at that.

RETIRING WITH A PENSION, THE BOB CLANTON STORY

The year was 1970, and Bob Clanton was about to graduate Northeastern University. With the company recruiters setting up shop near the downtown Boston campus, Bob put on his best suit to talk to a telephone company manager.

When he arrived for the interview, he was told; "Oh, that guy is busy, why don't you go talk to this guy." Bob said; "Sure, thinking, hey, I've got a suit on, what do I care?"

During the interview, he was given a problem to solve. As he recalls, the problem was something about a real estate company having telephone communications problems. His solution had something to do with putting a speaker phone in a conference room. To this day, he doesn't know where the inspiration came from, but the manager apparently thought it was brilliant.

All during the rest of the interview, all Bob could think about was if he'd have enough money to get his car out of the parking garage. Nevertheless, he was hired, but he had no idea it was for a sales position with New England Telephone.

He began by selling the series 300 switchboards. He soon graduated to large accounts, and then major accounts. Over the years, he has stayed with the same company, but like the experience of many in our industry, the company has changed. When the Bell system was broken up on January 1, 1984; he went with the American Bell group, headed by Arch McGill, in 1966 it merged back into AT&T; AT&T spun off Lucent, that recently was spun off as EXP@NETS.

Over the years, Bob has held positions as Account Executive, Product Manger, Product Marketing Manager, Administrative Manager, Sales Support Manager, and Channel Marketing Manager, but most of the time he's been a Sales Manager.

As Lucent finalizes the terms and conditions of the EXP@NETS spin-off, Bob finds himself in the "over 75" category. This is any combination of age and years of service that add up to 75. At age 52 with 30 years of service he easily qualified for retirement and a pension. In fact, he must "retire" from Lucent. Without revealing the size of his pension, I can tell you that in most years, I've worked three or four months to get what he'll get before he even gets out of bed in the morning. Sounds like a pretty good definition of success to me. Not to Bob.

Of course, he feels successful, and is wildly proud of what he's accomplished. I've know Bob for many years, we've worked a lot together, since he would hire me to speak to his salespeople and do sales and presentations skills training. He is highly respected by his peers, by the people that work for him, and by those that work with him.

I first met Bob when he called to buy one of my audio cassette competitive sales training tapes. At the time, I was marketing a tape set that showed salespeople how to "beat" the AT&T products that Bob was selling. He wanted to know what I was saying about his products, so he could develop counter sales strategies. Only much later did I begin to realize that the AT&T managers like Bob, who called me, were the brightest and best at what they did. Bob Clanton has success written all over him, and while he "must" retire, he feels like he's right in the middle of his journey.

FAILURE HITS THE AUTHOR IN THE FACE

Here's something I can assure you will happen in your life. Just when things are going the best, just when you begin to think that life is coming easy to you, just when you're "cooking", as we say; failure will raise it's ugly head. This very year, I was humbled. Humbled right to the very core of my being.

Many of you know that I'm a certified instructor for the prestigious Holden Corporation. Holden's clients read like the who's who of industry. Holden is positioned beautifully at the high end of the sales training industry. Their methodologies, documentation, support, and sales processes are terrific, and I'm proud to be associated with them. Over the past five years, approximately 20% of my revenues have been consistently generated through my relationship with Holden.

Recently, I had a huge setback. Here's the story. Holden had a new worldwide client. I was asked if I wanted to be on the facilitation team. We would be conducting the usual 3-day sales training course, and adding a fourth day, a sales automation day. To learn the custom sales automation software well enough to train the client's salespeople was going to take a sizable commitment of my time. Nevertheless, based on the size of the client, the potential of future business, and a chance to be on this Holden team, I took the time to learn the material.

After conducting just one class for the client, and exactly one and a half days before I was scheduled to leave for Sydney, Australia and Singapore to conduct more sessions, the customer told my contacts at Holden, that they "didn't want me on the team."

I was stunned. At dinner that night with my Holden friends, I was speechless. (Doesn't happen too often to this "speaker by trade".) It was over, the decision had been made. I was to be replaced.

Now I know that this story will bring back awful memories to some of you. When something like this happens, you're angry and hurt. Your pride has been given a blow, and no matter what level in life you are, your self-esteem takes a beating. You want an explanation, you want to fight back, you think about what you could have done differently.

I began to weigh my alternatives. The damages were severe. Not only had I already lost time and out of pocket money preparing for this assignment, but now I had to scramble to replace this future stream of cash flow with other clients. I could have pursued legal channels, as my damages were real. I could have gone directly to the client, around Holden, to get a full explanation, argued my case, and maybe worked something out.

Either approach would have put my Holden relationship in jeopardy. It's a relationship that I cherish. In fact, as it turned out, a couple of senior Holden executives supported me, went out on a limb for me and argued on my behalf. This actually encouraged me, because at first I thought the setback might impact the whole Holden relationship.

In the end, I got a fair answer, yes, politics were involved, yes, personalities were involved, and yes, I needed to improve in some areas. Am I still a worthwhile, talented, human being? Of course I am. Today, the Holden relationship in still intact, thriving, and I'm moving forward. Am I a failure? Of course I am not. Did I feel like a failure for awhile? Of course I did. Did I learn a lot from this setback? Of course I did. How long did I dwell on it? Just long enough to figure out what I should have done differently, and go on from there.

So what do the three running buddies, my barber, Bob Clanton, and the author have in common?

We've all had our bumps in the road. I can assure you that high up on that mountain in far off Equador, Rich, Rick, and John really wanted to climb that last 200 feet. I can assure you that when the interest rates climbed to over 20 per cent, or when the credit markets were the tightest, or when the Boston real estate market was falling, then my barber must have wondered what he was doing. I am sure that there were times in Bob Clanton's 30-year telecommunications career, when he'd had it with his company.

And if any of you, as you read this final chapter, dream of writing your own book someday, give me a call, and I'll share with you how many times you can anticipate calling a publisher before they'll print your book. Think of those calls as just a short stop on your journey of success, you'll be OK.